STUDIES IN GEOPHYSICS

Estuaries, Geophysics, and the Environment

Geophysics of Estuaries Panel
Geophysics Study Committee
Geophysics Research Board
Assembly of Mathematical and Physical Sciences
National Research Council

NATIONAL ACADEMY OF SCIENCES
Washington, D.C.   1977

NOTICE: The project that is the subject of this report was approved by the Governing Board of the National Research Council, whose members are drawn from the Councils of the National Academy of Sciences, the National Academy of Engineering, and the Institute of Medicine. The members of the Committee responsible for this report were chosen for their special competences and with regard for appropriate balance.

This report has been reviewed by a group other than the authors according to procedures approved by a Report Review Committee consisting of members of the National Academy of Sciences, the National Academy of Engineering, and the Institute of Medicine.

The Geophysics Study Committee is pleased to acknowledge the support of the National Science Foundation, the U.S. Geological Survey, the Energy Research and Development Administration, the National Oceanographic and Atmospheric Administration, the Defense Advanced Research Projects Agency, and the National Aeronautics and Space Administration for the conduct of this study.

Library of Congress Catalog Card Number 77-81272

International Standard Book Number 0-309-02629-6

Available from:

Printing and Publishing Office, National Academy of Sciences
2101 Constitution Avenue, Washington, D.C. 20418

Printed in the United States of America

Geophysics of Estuaries Panel

CHARLES B. OFFICER, Dartmouth College, *Chairman*
KENNETH F. BOWDEN, University of Liverpool
GABRIEL T. CSANADY, Woods Hole Oceanographic Institution
DOMINIC M. Di TORO, Manhattan College
KEITH R. DYER, Institute of Oceanographic Sciences (U.K.)
RICHARD W. GARVINE, University of Connecticut
RONALD J. GIBBS, University of Delaware
DOUGLAS E. HAMMOND, University of Southern California
DONALD R. F. HARLEMAN, Massachusetts Institute of Technology
ROBERT R. KIRBY, Institute of Oceanographic Sciences (U.K.)
DONALD J. O'CONNOR, Manhattan College
CURTIS R. OLSEN, Lamont-Doherty Geological Observatory
W. R. PARKER, Institute of Oceanographic Sciences (U.K.)
MAURICE RATTRAY, JR., University of Washington
H. JAMES SIMPSON, Lamont-Doherty Geological Observatory
ROBERT V. THOMANN, Manhattan College
KARL K. TUREKIAN, Yale University
SUSAN C. WILLIAMS, Lederle Laboratories

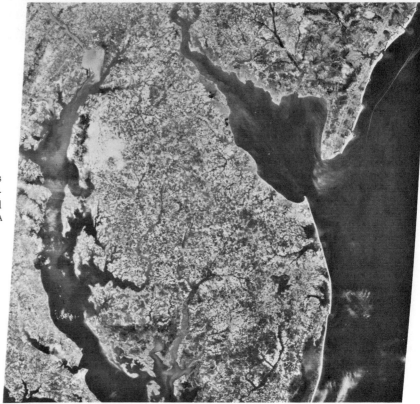

Delaware Bay and northern three quarters of Chesapeake Bay showing turbidity maximum in the upper Chesapeake Bay and complex distribution in both bays. (NASA ERTS imagery)

Delta of the Mississippi River and plumes of suspended material carried seaward. (NASA ERTS imagery)

Geophysics
Study Committee

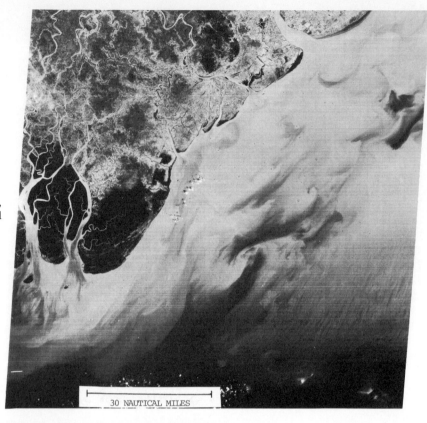

Mouths of the Irrawady River, Burma, showing complex pattern of suspended material offshore. (NASA ERTS imagery)

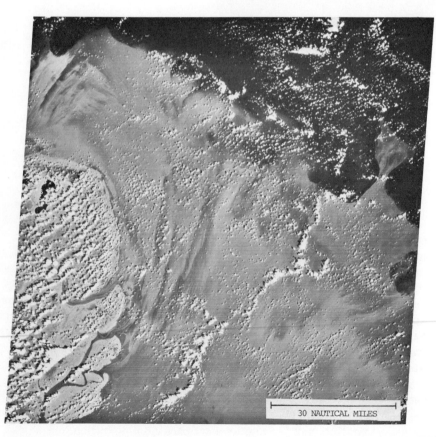

Ocean area near the mouth of the Amazon River (lower left). The figure shows tidal streaking and plumes extending more than 75 nautical miles offshore. (NASA ERTS imagery)

Preface

Early in 1974, the Geophysics Research Board completed a plan, subsequently approved by the Committee on Science and Public Policy of the National Academy of Sciences, for a series of studies to be carried out on various subjects related to geophysics. The Geophysics Study Committee was established to provide guidance in the conduct of the studies.

One purpose of the studies is to provide assessments from the scientific community to aid policymakers in decisions on societal problems that involve geophysics. An important part of such an assessment is an evaluation of the adequacy of present geophysical knowledge and the appropriateness of present research programs to provide information required for those decisions. When appropriate, the implications of this evaluation for research strategy may be set forth. Some of the studies place more emphasis on assessing the present status of a field of geophysics and identifying the most promising directions for future research. Topics of studies for which reports are currently in preparation include geophysical predictions, upper-atmosphere geophysics, energy and climate, water and climate, and estuaries.

Each study is developed through meetings of the panel of authors and presentation of papers at a suitable public forum that provides an opportunity for discussion. In completing final drafts of their papers, the authors have the benefit of this discussion as well as the comments of selected scientific referees. Responsibility for the individual essays rests with the corresponding authors.

The essays in this volume were presented in preliminary form at an American Association for the Advancement of Science meeting that took place in Boston, Massachusetts, in Febuary 1976. The symposium was divided into two sessions of six papers each: (1) Geophysical Investigations, chaired by Donald W. Pritchard, and (2) Engineering and Environmental Investigations, chaired by Bostwick H. Ketchum.

The introductory chapter provides an overview of the study, summarizing the highlights of the essays and formulating conclusions and recommendations. In preparing it, the Chairman of the Panel had the benefit of meetings and discussions that took place at the symposium and the comments of the Panel of authors and selected referees. Responsibility for its content rests with the Geophysics Study Committee and the Chairman of the Panel.

Contents

Estuaries, Geophysics, and the Environment

Overview
and
Recommendations

INTRODUCTION

Historically, the term "estuary" has been applied to the lower tidal reaches of a river. According to a contemporary definition, an estuary is a semienclosed body of water having a free connection with the open sea and within which the seawater is measurably diluted with freshwater drained from the land. This broader definition may include bays, sounds, inlets, fjords, and lagoons.[1] Virtually any human activity that affects the flow and contents of rivers may affect an estuary.

Estuaries have a historical and continuing importance to human activities. In view of the abundance of life in estuaries, the protection for vessels, and the natural flushing action for wastes, it is not surprising that early settlements in America were concentrated on estuaries or that one third of the present U.S. population lives and works close to them. Two thirds of the larger cities of the world lie on or near estuaries.

In many regions, estuarine environments are now being subjected to stress by activities of man: river flows are altered, estuaries are dredged and tidal flats reclaimed, and estuarine waters are used for effluent dispersal and industrial cooling. Such stresses are likely to become more intense in the future. Until recently, little thought was given to the ecological consequences. There is now a growing consciousness of the need to protect the estuarine environment.

1

In the United States, billions of dollars are spent each year in the operation of effluent control facilities.[2] Billions more are being spent on construction of facilities designed to stem the flow of pollutants into our estuarine and coastal waters. Hundreds of millions are being spent each year for surveys assessing the environmental impact of present and projected facilities. Comparatively little is being spent on research devoted to an understanding of the ecosystems that these activities seek to save—on understanding the chemistry, biology, and geophysics of the estuarine environment.

The principal reason for this imbalance relates to responsibility. The construction and operation of effluent control facilities is the responsibility of various local, state, and federal authorities. Evironmental impact surveys are a necessary consequence of new state and federal environmental protection laws. No single federal agency has the primary responsibility for the related environmental research.

The Federal Water Pollution Control Act has focused attention on the estuarine environment. The objective of that Act is "to restore and maintain the chemical, physical, and biological integrity of the Nation's waters" with goals for 1977, 1983, and 1985.[3] The Act established a National Study Commission whose responsibilities include making a complete investigation and study of all aspects of the economic, social, and environmental effects of achieving or not achieving the goals set for 1983.[4] Although some effects are dramatically evident, most are not. A thorough understanding of effects—especially long-term effects—requires a basic understanding of estuarine processes. The concept of estuarine management is evolving.[5] Effective estuarine management will be dependent on this basic understanding.

Compared with the science of rivers and lakes on the one hand and that of deep oceans on the other, the science of estuaries is extraordinarily complex. A number of disciplines are involved in understanding estuaries. If we consider strictly the geophysics of water bodies, there is a recognizable group of scientists—hydrologists—studying the inland waters and there is a similar group—deep-sea oceanographers—studying the oceans; there is not an analogous, identifiable group studying estuaries and their associated coastal waters.

Because circulation and mixing patterns dominate almost all other aspects of estuarine behavior, hydrodynamics is fundamental to all estuarine science. Circulation and mixing patterns are difficult both to measure and to analyze because they do not settle into a steady state but undergo ceaseless change in response to the ebb and flood of the tides, variations in freshwater flow, erratic effects of winds, irregularities in the estuarine geometry, and density differences between freshwater and salt water.

There is a wide range of scientific and engineering interest in estuaries[6] as evidenced, for example, by the publication of numerous reports[7] and conferences.[8] The 12 papers in this volume deal with many aspects of estuarine science, with emphasis on the role of geophysics. They do not treat such aspects as marine food harvests or navigational dredging but deal with phenomena basic to these and other applications. Their scope includes not only geophysical processes, including hydrodynamical and geological processes, but also basic chemical and biological phenomena, especially those related to water quality.

CONCLUSIONS AND RECOMMENDATION

Almost all concerns about estuaries relate to matters that affect life, including man and man's sensibilities. Intelligent management decisions related to estuaries increasingly depend on a thorough scientific understanding of the estuarine ecosystem—an understanding that is now lacking. The costs for research toward a basic understanding of estuaries are small compared with the costs of construction and operation of facilities intended to protect the estuarine waters. They are also small in comparison with the costs of the associated environmental surveys and the potential waste resulting from incorrect decisions.

This study did not include efforts to propose specific research programs. It concluded that there are two general problems to be addressed and formulated one recommendation.

CONCLUSIONS

■ There is a wide range of federal, state, and local agencies that seek answers to pressing practical questions relating to estuaries. Most of the questions posed by these agencies cannot be properly answered without an understanding of fundamental aspects of estuarine behavior. Of these, mixing and circulation are dominant for most problems in the majority of estuaries.

■ Even in those governmental and nongovernmental organizations with considerable interest in estuarine problems, estuarine research tends not to be clearly identified in the main structure of the organization. This lack of a visible focal point for estuarine problems is a potential obstacle to the development of effective, long-range, interdisciplinary research programs.

RECOMMENDATION *An overall review of the national efforts in estuarine research should be conducted.* Questions addressed by this review should include the following:

■ Is there adequate attention to solution of the basic problems in estuarine behavior?

■ Do the research programs have the kind of continuity that can reasonably be expected to deliver solutions to problems that require long-range research strategy?

■ Can steps be taken in research organizations concerned with estuarine problems to provide an improved focus for clear identification of estuarine research activities, for coordinated planning, and for effective interaction among the various scientific fields required to deal with the complex problems fundamental to estuarine research?

Because of the wide range of agencies concerned with estuaries, this review could appropriately be carried out under the auspices of the Federal Coordinating Council for Science, Engineering, and Technology. Estuarine problems potentially belong under the cognizance of at least two of its problem-oriented committees—the Committee on the Atmosphere and Oceans and the Committee on Energy and Natural Resources.

GEOPHYSICAL ASPECTS

Salinity measurements provide a useful basis for the classification of estuaries. *Well-mixed* estuaries have little variation of salinity with depth and are usually shallow with slight river current and strong tides. *Stratified* estuaries are usually deep and without strong tidal currents, so that freshwater flows to the sea over an underlying salt-water layer. Fjords are a special case of stratified estuaries primarily because of their great depth relative to width and the common occurrence of a shallow sill at the mouth. In such estuaries, a zone of high salinity gradient, called a halocline, separates the upper and lower portions of the water column. When the river is narrow and the flow is great, freshwater may fill the entire mouth, causing the brackish river water to form a distinguishable "plume" that floats out to sea over the salt water. The Amazon plume covers about a million square miles of ocean; the Connecticut River plume about 50 square miles of Long

3

Island Sound. In a river with large flow and a deep mouth (e.g., certain distributaries of the Mississippi delta), the denser seawater may intrude landward under the freshwater but be held back as a stable salt wedge underlying a strong freshwater flow above.

The objective of the estuarine hydrodynamicist is to devise theoretical and numerical models of the circulation and mixing processes that are consistent with observation and that permit prediction of hydrodynamic behavior under varying conditions. To do this, it is necessary to make judicious approximations to separate the variables of interest into manageable relationships. Among these are averaging processes—for example, an average over the tidal cycle or over the estuarine cross section. River flow in an estuary is superimposed on the tidal currents, which may be much larger. However, for many purposes tides may be regarded as simple oscillations. Similarly, the horizontal density gradient in stratified and in well-mixed estuaries gives rise to flow that varies with depth—with a seaward freshwater component above and a landward salt-water component below, superimposed on both tides and river current. When lateral circulation is not significant, two-dimensional models can give excellent agreement with experiment for velocity and salinity profiles, averaged over tidal cycles.

Where estuarine channels are curved or asymmetric, or wide enough to be significantly influenced by the earth's rotation (Coriolis effect), lateral flow may be superimposed on the longitudinal, giving rise to a spiral motion of the current. This can strongly affect some processes such as sedimentation, mixing, or dispersal of pollutants. Because the lateral flow component is usually one tenth or less of the longitudinal flow, accurate measurement is difficult.

Turbulence—of special importance in the transport and deposition of suspended particles—is an even more difficult hydrodynamic process to study in estuaries because it is small in scale, both in space and in time. All estuarine flow is turbulent, and when the turbulence reaches high intensity, high shear stresses result. In certain estuarine flows, it has been found that large shear stress exists during only a small portion of the tidal cycle. Because shear stresses determine rates of mixing and, near the bottom, rates of uptake of sediment, mean velocities and mean shear stresses might drastically understate the potential of a given flow for mixing and transport of suspended materials. The intensity and scale of turbulence are measured by flow meters of small size and short time constant. Extensive arrays of instruments are required to yield meaningful results. Theoretical analysis involves converting such data to suitable coefficients that describe the effect of turbulence on the flow.

ENGINEERING AND ENVIRONMENTAL ASPECTS

The last six papers in this volume extend the discussion into the areas of the chemistry and biology of the estuarine circulation and of the physically suspended particulate matter. They deal with naturally occurring constituents of the waters, as well as those produced by human activities, and their interactions and changes. Research involves measuring the various incoming constituents; tracing their chemical, biological, or physical changes until they reach the sea; and finally establishing what are called "biogeochemical models" of the estuary to permit prediction of the effect of changes.

Water quality and pollution have become subjects of increasing concern in recent years because of the effects of growing population and industrialization. Pollution has often closed beaches and fishing grounds, has directly killed estuarine aquatic life, and has damaged spawning grounds of estuarine and ocean fish. Detriments to water quality are of three major types: those directly harmful to marine life, to humans, or to both, such as dangerous bacteria or chemicals (e.g., Kepone, polychlorinated biphenyls, and heavy metal ions); those which in excess reduce the dissolved oxygen concentration (carbonaceous and nitrogenous compounds); and excess nutrients that stimulate algal growth (inorganic nitrogen and phosphorus compounds).

The constituents that have received most public attention are organic materials that consume dissolved oxygen as they undergo bacterial degradation, metal ions from both natural drainage and industrial wastes, nutrients of aquatic plant life (inorganic nitrogen compounds and phosphates), suspended particulate matter, and undesirable bacteria from sewage. Organic compounds that are both harmful and not biodegradable, such as Kepone and polychlorinated biphenyls (PCB'S), have received increasing attention recently as their dangers in riverine and estuarine waters have been recognized. The discharge of Kepone into the James River and PCB's into the Hudson River resulted in the closing of both rivers to commercial fishing and caused great anxiety regarding long-term effects. These chemicals are biologically dangerous and extremely stable. They can be concentrated in the marine food chain, including species used as food for humans.

The most common pollutants in all bodies of water are wastes that have *biochemical oxygen demand* (BOD)—mainly raw sewage or biodegradable industrial wastes. Oxygen-demanding materials and nutrients damage aquatic life directly as well as indirectly through a process known as eutrophication. Most organic compounds are biodegradable; such pollutants are of two main classes, the carbonaceous and nitrogenous. Both are utilized by bacteria, with the carbon content being converted to carbon dioxide and the nitrogen to ammonia or other inorganic nitrogen compounds that are nutrients for algal growth. In these processes, the amount of dissolved oxygen of the river or estuary is depleted; if the BOD waste level is high enough, marine life cannot successfully compete for the oxygen (for example, in the upper portion of the Houston Ship Canal, the dissolved oxygen concentration is zero). Excess BOD wastes may be kept out of estuaries by diverting sewage flows to other discharge sites or by treatment. Nitrogen and phosphorus compounds may also be substantially removed by suitable treatment at added expense. They are beneficial in proper concentration but in excess bring about so much plant growth that eutrophication results.

Even a modest improvement in understanding the effects of such materials can lead to major engineering and economic payoffs. For example, it is probable that a better understanding of this aspect of estuarine science could lead to conclusions that, in some cases, particular treatment processes are unnecessary and, in others, ocean outfalls could be a more satisfactory solution than additional treatment facilities. These decisions involve large sums; they warrant an adequate scientific basis.

Various types of materials can be used as chemical tracers in the study of estuarine processes. A rich resource for this purpose is the variety of metal ions present in trace amounts, both from natural sources and industrial wastes, which undergo important and informative changes in the tidal mixing region. For example, the concentrations of cobalt, nickel, and silver brought down the Housatonic River (Connecticut) show a sharp rise in the mixing zone, followed by a decline to very low values in the salt water in Long Island Sound. It is presumed that these ions are absorbed by the salt-water effect and are later precipitated by other reactions. It is also possible that they may be involved in oxidation–reduction reactions of the tidal marshes. Understanding this chemistry better is important in connection with decisions related to such wastes—their effects, the amounts that can be safely accommodated without damage to estuarine ecology, and their disposition.

5

A role of estuarine scientists in water-quality engineering is to formulate models that take into account not only hydrodynamics but also biological and chemical processes. Such theoretical models involve a description of the incoming constituents, their chemical and biological transformations, the role of circulation in such changes, and the effects of the variation of the different constituents and conditions. These models become extremely complex; to make useful progress, attention must be focused on the critical variables. Even in nitrogen-limited ecosystems, such as the Potomac and Pamlico Estuaries, this is not simple because there are at least seven types of nitrogen compound involved, as well as many chemical and biological transformations.

It is unrealistic to strive to achieve improved water quality in estuaries by halting all discharges generated by human activity. Instead, well-informed management of the various uses of estuaries can prevent harm and even bring benefits. For example, nitrogen and phosphorus nutrients discharged in proper amounts at the right places might benefit aquatic life just as they benefit farm crops. Excess of one nutrient may not be harmful when algal growth rate is controlled by another factor such as water temperature, turbidity, and sunlight. If excess phosphate could be accommodated, costly tertiary waste treatment required to remove it could be avoided.

RESEARCH STRATEGY

With the increasing problems of water supply, and increased concern about the well-being of the environment for aesthetic and practical reasons, there will be increasing attention to problems that must be solved by estuarine science. They may seem initially to be purely engineering, biological, chemical, or geological problems, but, as the posers of the questions seek answers, they will discover that the solutions often depend on geophysical considerations.

High-quality estuarine research programs do exist, but most have been oriented toward specific problems—mainly marine food harvests, maintenance of navigational channels, and water quality. What is neglected is the basic interdisciplinary research, involving a wide range of scientific fields, that would be beneficial and fundamental to all programs concerned with protecting the estuarine environment.

Many federal agencies are directly concerned with estuarine science. With the exception of the National Science Foundation, all the agencies are concerned as a result of practical considerations related to their missions. Federal organizations concerned with estuaries include the Department of Commerce (National Oceanic and Atmospheric Administration, National Marine Fisheries Service), Department of the Interior (U.S. Geological Survey, Bureau of Land Management, Park Service, Fish and Wildlife Service, Bureau of Reclamation), National Aeronautics and Space Administration, Environmental Protection Agency, Energy Research and Development Administration, Nuclear Regulatory Commission, Department of Defense (Corps of Engineers, Navy), and Department of Transportation (Coast Guard). Numerous state and local agencies are concerned with estuaries. At one time or another, all these agencies have addressed the geophysical problems of estuaries and coastal waters.

Government laboratories necessarily direct their research efforts toward practical problems related to their missions. University and academic research laboratories have a tradition of commitment to basic research—over long periods of time, if necessary—without the requirement of a foreseen applied result. Government laboratories in general are more actively concerned with transferring the results of basic research to applications. Scientific and technical talent capable of dealing with the difficult problems of estuarine science can be found in both government and academic institutions. A long-range research strategy should take account of an appropriate balance among these considerations.

Basic research on estuaries must involve the talents of a wide range of specialists. Many of the scientists and engineers who could make effective contributions are not at present directly involved with estuarine problems.

Variability among estuaries makes generalizations from one estuary to another a risky undertaking. A few estuaries have been studied intensively; most have not. The benefits of new understandings would be greatly multiplied if it were not necessary to study each estuary independently. Therefore, research strategy should include efforts to determine whether there are key parameters for given types of estuarine conditions that could be used to produce models of even limited transferability of understanding from one estuary to another.

While the concerns about the environment have given rise to numerous *ad hoc* investigations and surveys, there has been no program of continuing research involving multidisciplinary approaches designed to provide solutions to fundamental questions of estuarine science that must be understood for proper assessment of specific problems. Many of the most important problems are not subject to short-term, project-type solutions; long-term continuity and breadth of viewpoint are essential.

NOTES

1. This definition of estuaries was formulated by D. W. Pritchard and published with considerable discussion in *Estuaries,* Publication 83 of the American Association for the Advancement of Science, Washington, D.C., 1967. Public Law 92-500 (October 18, 1972) entitled "Federal Water Pollution Control Act Amendments of 1972" contains the following definition in Section 104 (n)(4):

   For the purpose of this subsection, the term "estuarine zones" means an environmental system consisting of an estuary and those transitional areas which are consistently influenced or affected by water from an estuary such as, but not limited to, salt marshes, coastal and intertidal areas, bays, harbors, lagoons, inshore waters, and channels, and the term "estuary" means all or part of the mouth of a river or stream or other body of water having unimpaired natural connecton with open sea and within which the sea water is measurably diluted with fresh water derived from land drainage.

2. Pollution control expenditures. The following data are reproduced from *Environmental Quality,* Seventh Annual Report of the Council on Environmental Quality, September 1976 (Tables 1-35 and 1-37). These figures are indicative of the magnitude of expenditures to control water pollution. Directly or indirectly, much of this expenditure is related to estuarine pollution.

   Estimated Water Pollution Control Expenditures in 1975 ($ Billion)

	Incremental*	Total
Operation and maintenance	3.0	5.0
Capital	1.7	9.7
	4.7	14.7

   * Incremental costs are expenditures made pursuant to federal environmental legislation beyond those that would have been made in the absence of this legislation.

3. Public Law 92-500 (October 18, 1972) "Federal Water Pollution Control Act Amendments of 1972":

   Section 101. (a) The objective of this Act is to restore and maintain the chemical, physical, and biological integrity of the Nation's waters. In order to achieve this objective it is hereby declared that, consistent with the provisions of this Act—

   (1) it is the national goal that the discharge of pollutants into the navigable waters be eliminated by 1985;

   (2) it is the national goal that wherever attainable, an interim goal of water quality which provides for the protection and propagation of fish, shellfish, and wildlife and provides for recreation in and on the water be achieved by July 1, 1983;

   (3) it is the national policy that the discharge of toxic pollutants in toxic amounts be prohibited;

(4) it is the national policy that Federal financial assistance be provided to construct publicly owned waste treatment works;

(5) it is the national policy that areawide treatment management planning processes be developed and implemented to assure adequate control of sources of pollutants in each State; and

(6) it is the national policy that a major research and demonstration effort be made to develop technology necessary to eliminate the discharge of pollutants into the navigable waters, waters of the contiguous zone, and the oceans.

Section 102. (a) The Administrator shall, after careful investigation, and in cooperation with other Federal agencies, State water pollution control agencies, interstate agencies, and the municipalities and industries involved, prepare or develop comprehensive programs for preventing, reducing, or eliminating the pollution of the navigable waters and ground waters and improving the sanitary conditions of surface and underground waters. In the development of such comprehensive programs due regard shall be given to the improvements which are necessary to conserve such waters for the protection and propagation of fish and aquatic life and wildlife, recreational purposes, and the withdrawal of such waters for public water supply, agricultural, industrial, and other purposes.

Section 104. (a) The Administrator shall establish national programs for the prevention, reduction, and elimination of pollution and as part of such programs shall—

(1) in cooperation with other Federal, State, and local agencies, conduct and promote the coordination and acceleration of, research, investigations, experiments, training, demonstrations, surveys, and studies relating to the causes, effects, extent, prevention, reduction, and elimination of pollution.

4. Public Law 92-500 (October 18, 1972) "Federal Water Pollution Control Act Amendments of 1972":

Section 315. (a) There is established a National Study Commission, which shall make a full and complete investigation and study of all of the technological aspects of achieving, and all aspects of the total economic, social, and environmental effects of achieving or not achieving, the effluent limitations and goals set forth for 1983 in section 301 (b)(2) of this Act.

The Commission issued a "Staff Report to the National Commission on Water Quality," in April 1976.

5. See, for example, W. O. Spofford, Jr., C. S. Russel, and R. A. Kelly, *Environmental Quality Management: An Application to the Lower Delaware Valley,* Resources for the Future, Inc., Research Paper R-1, September 1976.

6. The range of interest in estuaries is evidenced by the attendance of some 450 participants in the Third International Estuarine Research Conference, Galveston, Texas, October 1975. They came from departments of oceanography, marine science, biology, zoology, botany, limnology, engineering, geology, and geosciences and from a diversity of academic, government (state and federal), and industrial organizations concerned with estuaries, wetlands, fisheries, and environmental science. See Footnote 8 for further discussion of the Estuarine Research Federation.

The Office of Water Research and Technology, Department of the Interior, has issued three volumes entitled *Estuarine Pollution—A Bibliography* totaling 1216 pages. They provide an indication of the extensive interest in pollution in the estuarine environment.

7. Studies and Reports. Estuaries have been given attention in many studies and reports, especially those concerned with coastal waters and pollution. The following is a small sampling of such reports:
Our Nation and The Sea: A Plan for National Action. Report of the Commission on Marine Science, Engineering and Resources, J. A. Stratton, chairman, January 1969. Establishment of the Commission was authorized by Public Law 89-454, June 1966. The transmittal of this report to the President and the Members of Congress stated:

In response to its mandate, the Commission has undertaken an intensive investigation of a broad array of marine problems ranging from the preservation of our coastal shores and estuaries to the more effective use of the vast resources that lie within and below the sea. The recommendations which have emerged from this study constitute a program which we believe will assure the advancement of a national capability in the oceans and go far towards meeting the inevitable needs of the future.

The report was supplemented by three volumes containing reports of eight panels, among them the Panels on Basic Science and on Management and Development of the Coastal Zone.

Regarding estuaries, the findings of these panels are in good agreement with the present report. The basic problems have not changed or been solved. The Panels' recommendations include the following:

The Nation should undertake a much enhanced program of basic research into the dynamics of estuarine waters, the identification of specific pollutants and the tracing of their effects, both on individual species and ecosystems, and on the mechanisms through which organisms in the estuarine ecosystem take up and accumulate various kinds of pollutants.

Man's Impact on the Global Environment: Assessment and Recommendations for Action. Report of the Study of Critical Environmental Problems (SCEP) sponsored by the Massachusetts Institute of Technology, C. L. Wilson and W. H. Matthews, eds., MIT Press, Cambridge, Mass., 1970. The need for this study was perceived in connection with governmental and nongovernmental preparations for the 1972 United Nations Conference on the Human Environment. The intent was to provide an authoritative assessment of the degree and nature of man's impact on the global environment with specific recommendations for new programs of focused research, monitoring, and action. The study included estuaries and several recommendations related to their pollution. It drew attention to the multiplicity of authorities concerned with river basins, estuaries, and coastal zones.

Marine Environmental Quality: Suggested Research Programs for Understanding Man's Effect on the Oceans, the report of a special study held under the auspices of the Ocean Science Committee of the NRC's Ocean Affairs Board, August 1971. This report emphasizes that the behavior and ultimate fate of pollutants in estuaries is poorly understood because of the complexities of the transport processes and the chemical reactions.

Recommendations for Basic Research on Transfer Processes in Continental and Coastal Waters, report of a workshop sponsored by the National Science Foundation held at Block Island, July 1974. This workshop was convened to review the current status of basic research in continental and coastal waters and "to define a focus for future research on processes involved in the movement of natural and anthropogenic materials through the aqueous environment of continents and their margins." The workshop noted that "while existing programs in environmental science may be establishing a valuable data base about the distribution of materials in selected areas, these programs provide little assistance in constructing a scientific framework to understand the processes which control these distributions."

The workshop emphasized the importance of circulation and mixing and stressed the fact that limited research programs aimed at specific problems "will do little to eliminate our ignorance of the underlying controlling mechanisms."

A Workshop to Identify Research Priorities in Estuarine Physical Transport Processes was held at the Marine Sciences Research Center, State University of New York, Stony Brook, New York, November 1976. The goals of the workshop were (1) to identify important unresolved problems of physical transport processes in estuaries; (2) to assess personnel and physical resources required for significant advancement; (3) to explore ways of encouraging interinstitutional cooperation; (4) to explore ways of increasing scientific and management effectiveness of estuarine monitoring studies. (The report of this workshop is in preparation.)

Perspectives on Technical Information for Environmental Protection, a Report to the U.S. Environmental Protection Agency from the Commission on Natural Resources and the Steering Committee for Analytical Studies, NRC, was issued in March 1977. This report (Volume I of the series) and its ten associated reports deal with the question of the role of science in environmental protection.

Part II of *Perspectives* contains an executive summary of four of the reports: Decision Making, Research, Monitoring, and Manpower. While not directed primarily toward estuaries, many aspects of these reports are pertinent to the estuarine environment. A basic theme of the report *Estuaries, Geophysics, and the Environment* appears also in *Perspectives* in the chapter "The Role of Scientific and Technical Information in Environmental Decision Making," which underscores the need for research "dedicated with continuity to long-range, fundamental investigation . . . necessary to understand ecosystems, to improve monitoring, and to develop a capability for anticipating environmental impacts."

8. *Conferences and Symposia.* There have been several conferences on estuaries in recent years.

The Symposium on Estuaries, Geophysics and the Environment held at the meetings of the American Association for the Advancement of Science, February 1976, was sponsored by the Geophysics Study Committee as an integral part of this study. The papers presented at that symposium are published as part of this volume.

Estuaries (referenced in Note 1) contains the proceedings of the Conference on Estuaries held at Jekyll Island, Georgia, 1964.

The Estuarine Research Federation, representing some 1200 members of four regional Estuarine Research Societies (Atlantic, New England, Gulf, and Southeastern), has organized biennial International Estuarine Research Conferences, beginning with a conference on Long Island in 1971. The society has arranged publication of the proceedings of the second and third conferences (held at Myrtle Beach, South Carolina, October 1973, and Galveston, Texas, October 1975, respectively):

Proceedings of Second Conference: *Estuarine Research,* L. E. Cronin, ed., Academic Press, New York, 1975. Vol. 1, Chemistry, Biology and the Estuarine System; Vol. 2, Geology and Engineering.

Proceedings of Third Conference: *Estuarine Processes,* M. Wiley, ed., Academic Press, New York, 1976. Vol. 1, Uses, Stresses and Adaptation to the Estuary; Vol. 2, Circulation, Sediments and Transfer of Materials in the Estuary.

A Symposium on Transport Processes in Estuarine Environments was held at the Belle W. Baruch Institute for Marine Biology and Coastal Research, University of South Carolina, in May 1976.

The Environmental Protection Agency sponsored a Conference on Estuary Pollution in Pensacola, Florida, February 1975. The proceedings of that conference were published: *Estuarine Pollution Control and Assessment, Proceedings of a Conference,* Vols. I and II, March 1977, Office of Water Planning and Standards, U.S. EPA.

I

GEOPHYSICAL
INVESTIGATIONS

Longitudinal Circulation
and Mixing Relations
in Estuaries

1

CHARLES B. OFFICER
Dartmouth College

INTRODUCTION

Our present understanding of the hydrodynamics of estuaries has evolved over the years through contributions from individuals with varying backgrounds in geophysics, physical oceanography, and civil engineering. Some of these contributions have come from purely scientific investigations, whereas others have been related to rather specific engineering problems. Recently there has been an increased emphasis in understanding the circulation and mixing processes in estuaries, and this has been directly related to the existing and potential future pollution problems of estuaries. Such an understanding is important in itself. In addition, considering the kinds of problems that currently confront investigators looking at the chemical and biological aspects of water quality and the geological aspects of sediment transport and distribution, a knowledge and understanding of estuarine circulation and mixing is essential.

We shall use here the definition of an *estuary* given by Pritchard (1952a) and Cameron and Pritchard (1965). This is that an estuary is a semienclosed body of water having a free connection with the open sea and within which the seawater is measurably diluted with freshwater derived from land drainage. Traditionally, the term "estuary" has been applied to the lower reaches of a river into which seawater intrudes and mixes with freshwater draining seaward from the land. The term has been extended to include bays, inlets, gulfs, and sounds into which several rivers empty and in which the mixing of freshwater and saltwater occurs.

Various attempts have been made to classify estuaries by type. Such a classification scheme for an estuary as a whole is difficult to do; for there are hardly any two estuaries that are exactly the same with regard to geometric and bathymetric configuration, the physical oceanographic characteristics of their circulation and mixing, or both. Further, any given estuary may show well-mixed or stratified conditions, for example, as a function of longitudinal distance along the estuary, season of the year, or even in some cases phase of the tidal cycle.

Nevertheless, a useful and convenient distinction can be

made in terms of the vertical salinity distribution. We define a *well-mixed* condition as one in which there is essentially no variation in the salinity in a vertical column. We can also distinguish a stratified condition with a halocline between the upper and lower portions of the water column of nearly constant salinity. We define a *weakly, or partially, stratified* condition as one in which there is a change in salinity of only a few parts per thousand from surface to bottom and a *strongly, or highly, stratified* condition as one in which there is a change in salinity of several parts per thousand from surface to bottom. We can also distinguish a condition in which there is an interface between two different water types. We define an *arrested salt wedge* as one in which there is a stable salt wedge underlying a strong, freshwater flow above and a *fjord entrainment* type flow as one in which there is a relatively stagnant deep water mass overlain by a thin river runoff flow.

LONGITUDINAL CIRCULATION AND MIXING

One of the better understood aspects of estuarine hydrodynamics is that of the longitudinal circulation and mixing characteristics for well-mixed and stratified estuaries as averaged over a tidal cycle. The driving forces for the circulation are the longitudinal surface slope, acting in a down-estuary direction, and the longitudinal density gradient force, related to the longitudinal salinity gradient, acting in an upestuary direction. These two driving forces are balanced by the internal and bottom frictional forces, and there may in some cases be an important contribution from a third driving force related to the wind stress at the surface. The surface slope force is constant as a function of depth, and the density gradient force increases essentially linearly as a function of depth. For the condition in which there is a small river runoff flow, which is common for many estuaries, the net effect is, then, that the surface slope force will be dominant in the upper portion of the water column producing a net circulation flow downestuary and that the density gradient force will be dominant in the lower portion of the water column producing a net circulation flow upestuary.

Some few years ago, Rattray and Hansen (1962) and Hansen and Rattray (1965) in two elegant papers developed solutions for this type of circulation and mixing, governed by the longitudinal density gradient, using similarity solution techniques. In their central regime for the estuary, they assumed that the longitudinal salinity gradient was constant as a function of depth. This has the effect of decoupling the motion and mixing equations so that each may be solved separately and directly rather than as a set of coupled partial differential equations. Such direct solutions have been given by Officer (1976) and have been extended to include bottom frictional effects and nonlinear tidal inertial and tidal bottom frictional effects.

For a simplified bottom boundary condition of zero velocity, the solution for the longitudinal circulation velocity, v_x, as averaged over a tidal cycle, is of the form

$$v_x = \frac{1}{48} \frac{g\lambda h^3}{\rho N_z} (1 + 9n^2 + 8n^3) + \frac{3}{2} v_0 (1 - n^2)$$

$$+ \frac{1}{4} \frac{hT}{\rho N_z} (1 - 4n + 3n^2), \quad (1.1)$$

where g is the gravitational acceleration constant, λ the longitudinal density gradient, h the water depth, ρ the water density, N_z the vertical eddy viscosity coefficient, assumed constant, and T the wind stress at the surface. The quantity $n = z/h$ is the fractional depth, and the quantity $v_0 = R/A$ is the sectionally averaged river runoff velocity, where R is the river flow and A the cross-sectional area normal to the flow. In this expression, the first term may be considered the density-gradient contribution, the second term the river-runoff contribution, and the third term the wind-stress contribution.

For a more preferred bottom boundary condition of a quadratic velocity relation of the form $k|v_x|v_x = -\rho N_z(\partial v_x/\partial z)$, where k is the bottom friction coefficient, the solution is of the form

$$v_x = \frac{gi}{2N_z}(h^2 - z^2) - \frac{g\lambda}{6\rho N_z}(h^3 - z^3)$$

$$- \sqrt{\frac{gh}{k} \left(\frac{1}{2} \frac{\lambda h}{\rho} - i \right)} \quad (1.2)$$

with the omission of the wind-stress term. We also now have the subsidiary relation

$$v_0 = \frac{gi}{3N_z} h^2 - \frac{g\lambda}{8\rho N_z} h^3 - \sqrt{\frac{gh}{k} \left(\frac{1}{2} \frac{\lambda h}{\rho} - i \right)} \quad (1.3)$$

for the surface slope, i, in terms of the longitudinal density gradient, λ, and the river-runoff velocity, v_0. We can rewrite these two expressions in terms of the velocity at the surface, v_s, and the velocity at the bottom, v_b, as

$$v_x = v_s(1 - 9n^2 + 8n^3) + v_b(-3n^2 + 4n^3)$$

$$+ v_0(12n^2 - 12n^3), \quad (1.4)$$

where

$$v_s = \frac{gh^2}{6N_z} \left(i - \frac{1}{4} \frac{\lambda h}{\rho} \right) + v_0 \quad (1.5)$$

and

$$v_b = -\frac{gh^2}{3N_z} \left(i - \frac{3}{8} \frac{\lambda h}{\rho} \right) + v_0. \quad (1.6)$$

For the salt continuity, or mixing, equation the essential balance is that between the longitudinal circulation advection flux and the vertical diffusion flux. The simplified form of the vertical salinity variation is, then, the double integral with respect to depth of the net circulation velocity distribution. For the velocity function of the form of Eq. (1.4) this is

$$s_1 = \frac{h^2}{K_z} \frac{\partial s}{\partial x} \left[v_s \left(\frac{2}{5} n^5 - \frac{3}{4} n^4 + \frac{1}{2} n^2 - \frac{1}{12} \right) \right.$$

$$+ v_b \left(\frac{1}{5} n^5 - \frac{1}{4} n^4 + \frac{1}{60} \right)$$

$$\left. + v_0 \left(-\frac{3}{5} n^5 + n^4 - \frac{1}{2} n^2 + \frac{1}{15} \right) \right], \quad (1.7)$$

where s_1 is the salinity deviation from its depth averaged mean, K_z is the vertical eddy-diffusion coefficient, assumed constant, and $\partial s/\partial x$ is the longitudinal salinity gradient.

We are interested here in comparing these derived relations with actual physical observations in estuaries. The relations provide us with a means not only of comparing field data with these simplified theoretical considerations but also of determining values for the parameters N_z, i, k, and K_z from the observations of v_s, v_b, v_0, $\partial s/\partial x$, h, and $s_b - s_s$, the salinity difference from the surface to the bottom. For this latter purpose we have from Eqs. (1.5) and (1.6)

$$N_z = \frac{1}{24} \frac{g\lambda h^3}{(2v_s + v_b - 3v_0)\rho} \quad (1.8)$$

and

$$i = \frac{3v_s + v_b - 4v_0}{8v_s + 4v_b - 12v_0} \frac{\lambda h}{\rho}, \quad (1.9)$$

from Eq. (3)

$$k = \frac{g\lambda h^2}{v_b^2 \rho} \left(\frac{1}{2} - \frac{3v_s + v_b - 4v_0}{8v_s + 4v_b - 12v_0} \right), \quad (1.10)$$

and from Eq. (7)

$$K_z = \frac{1}{20} \frac{(3v_s - v_b - 2v_0)h^2}{s_b - s_s} \frac{\partial s}{\partial x}. \quad (1.11)$$

There are only a few sets of sufficiently detailed physical observations in estuaries that permit such a comparison to be made. We shall examine data from two areas that can be described as well mixed to weakly stratified, Long Island Sound and the Mersey Estuary; two areas that can be described as stratified, the Southampton and James Estuaries; and two areas that can be described as strongly stratified, the Columbia River and Vellar Estuaries. In five of the six examples, river runoff is such that v_0 is considerably smaller than v_s or v_b. For all cases, for lack of any specific information to the contrary, the wind stress is assumed to be zero. It is important to restate two important assumptions in the derivations, specifically that N_z and K_z are constants and that there are no lateral effects. These conditions are not always met or approached in actual estuaries.

The James Estuary is tributary to Chesapeake Bay on the east coast of the United States. The field data are from Pritchard (1952b, 1954). Typical circulation velocity versus depth and mean salinity versus depth curves from the field data, along with theoretical curves obtained from the relations (1.4) and (1.7), are shown in Figure 1.1. The agreement is reasonable. The computed values for N_z, K_z, k, and i are given below in Table 1.2. Rattray and Hansen (1962) made a similar comparison with the James Estuary data, obtaining values of $N_z = 4$ cm^2 sec^{-1} and $K_z = 2$ cm^2 sec^{-1}, which are equivalent to the values obtained here. The first derivative of the field data circulation velocity does not approach zero at the surface, which may be indicative of a downestuary wind stress at the surface at the time of the measurements; and a better theoretical fit to the data could be obtained by including a wind-stress term. In the vicinity of the middepth halocline, the theoretical salinity curve shows less curvature than do the field data, indicative of lowered K_z values there, which is presumably related to a Richardson number effect in this vicinity. Pritchard (1967), using the salt continuity equation directly in sequential, finite difference form, calculated values of around $K_z = 6$ cm^2 sec^{-1} in the upper portion of the water column, $K_z = 1$-2 cm^2 sec^{-1} in the vicinity of the halocline, and $K_z = 4$ cm^2 sec^{-1} in the lower portion of the water column.

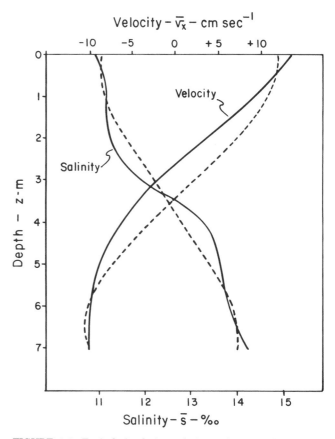

FIGURE 1.1 Typical circulation velocity and mean salinity versus depth curves for the James Estuary. Observations,—; theory,---. From Pritchard (1954).

The Mersey Estuary is located on the west coast of Great Britain. The field data are from Bowden (1960, 1962, 1963). The circulation velocity versus depth and the mean salinity versus depth curves from the field data for four measurement periods are shown in the upper and lower portions of Figure 1.2, respectively, along with the theoretical comparisons; the theoretical salinity comparison is for the data of June 30 to July 2, 1959, only. The agreement is reasonable. Again, the first derivative of the field data circulation velocities does not approach zero at the surface. Further, the field data mean salinities show lower values in the upper portion of the water column than would be anticipated from the simplified theory, corresponding perhaps to the higher downestuary circulation velocities in this region. Bowden (1960, 1962, 1963) also calculated values of N_z and K_z as a function of depth directly from the motion and salt continuity equations in finite-difference form. The values showed a substantial variation as a function of depth and from one measurement period to the next. The averaged values are $N_z = 39$ cm^2 sec^{-1} and $K_z = 22$ cm^2 sec^{-1}, which are in good agreement with the values obtained here.

The Southampton Estuary is located on the south coast of Great Britain. Its circulation characteristics are not so simple as those for the James or Mersey Estuaries; there are indeed substantial lateral effects there. The field data

are from Dyer (1973). The mean salinity versus depth curves and the circulation velocity versus depth curves are shown in the upper and lower portions of Figure 1.3, respectively, for four measurement periods at one station. Only one of the stations, that of April 6, 1966, shows what one might consider a typical longitudinal velocity profile. There were similar variable results at other stations, and the observed salinity distribution showed a considerable gradient across the estuary. Assuming that the data of April 6, 1966, were indicative of a condition for which the longitudinal circulation was dominant, the calculated values given in Table 1.2 and the comparison given in Figure 1.3 were obtained. The assumption is probably not justified, but at least it permits some parameter values to be calculated for this case.

The Vellar Estuary is located on the east coast of India. It is in an area of the monsoons, and the river-runoff rate can show large variations; for the measurement periods, the estuary condition was highly stratified. The field data are from Dyer and Ramamoothy (1969). The mean salinity and circulation velocity versus depth curves for four measurement periods at one station are shown in Figure 1.4. The river discharge rates for the four periods were 375, 90, 86, and 3 m^3 sec^{-1}, respectively. A comparison has been made for the data of January 27 and February 9, with the results as shown in Figure 1.4 and the computed parameter values as given in Table 1.2.

The Columbia River Estuary is located on the west coast of the United States. It is characterized by an unusual combination of large river discharge and strong tidal currents. The field data are from O'Brien (1952) and Hansen (1965) based on measurements taken by the Corps of Engineers in 1932. O'Brien (1952) made a separation of the river advection and the density gradient contributions to the circulation velocity, v_0, at a high river discharge period from which the values of N_z, i, and k have been calculated; and Hansen (1965) presented the mean salinity distribution as a function of depth from which, using the relation from Eq.

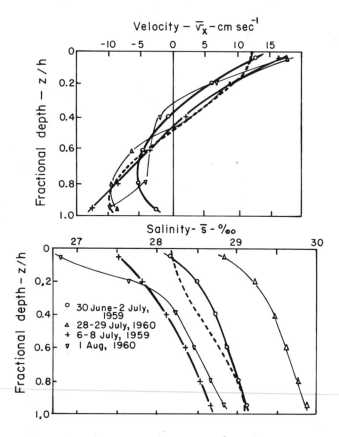

FIGURE 1.2 Circulation velocity and mean salinity versus depth curves for the Mersey Estuary. Observations,—, theory,---. From Bowden (1963).

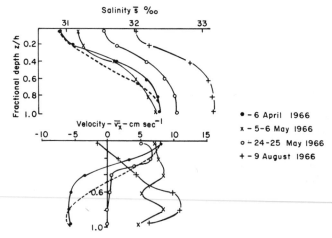

FIGURE 1.3 Circulation velocity and mean salinity versus depth curves for the Southampton Estuary. Observations,— theory,---. From Dyer (1973).

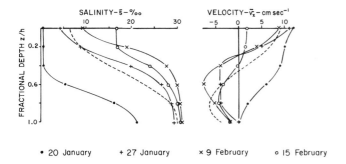

FIGURE 1.4 Circulation velocity and mean salinity versus depth curves for the Vellar Estuary. Observations,—; theory,---. From Dyer and Ramamoothy (1969).

(1.11), a value for K_z has been determined. Hansen and Rattray (1965) estimated values of $N_z = 25$ cm^2 sec^{-1} and $K_z = 15$–30 cm^2 sec^{-1} for this same case, which are somewhat higher than the N_z and K_z values obtained here.

Long Island Sound is a rather large, elongated, and estuary-like body of water between Long Island and the mainland on the east coast of the United States. The field data are from Riley (1952, 1955a, 1955b). Riley (1955b) includes a graph of the maximum ebb and flood tidal currents as a function of depth for two measurement periods at a single location. From these data a crude estimate of the circulation velocity as a function of depth can be made from which, along with the other data given, the values shown in Table 1.2 have been calculated.

Let us now look at the observed and computed values tabulated in Tables 1.1 and 1.2. From Table 1.1 we see from the $s_b - s_s$ values that a wide range of conditions has been covered from nearly well mixed for Long Island Sound and the Mersey Estuary, to stratified for the Southampton and James Estuaries, to strongly stratified for the Columbia River and Vellar Estuaries. We also see that the longitudinal salinity gradient, $\partial s/\partial x$, increases in a manner similar to the vertical salinity deviation, $s_b - s_s$, from one estuary to the next. The values of the bottom velocity, v_b, or in the case of the Columbia River Estuary, $v_b - v_0$, are negative in each case and are a substantial fraction of the corresponding surface velocity.

When we come to consider the viscosity and diffusion coefficients of Table 1.2, there are a number of generalities that should be kept in mind. We know in a qualitative manner, unfortunately without being able to be too quantitative about the description, that N_z and K_z will be scale, or depth, dependent; that in estuaries they will in some manner be related to the magnitude of the tidal action, which is the ultimate driving force for the turbulence phenomena observed there; and that they will decrease with increasing Richardson number or, what is the same, with an increasing ratio of the potential energy increase due to vertical mixing to the available kinetic energy of the turbulence. We also know from observations that the ratio K_z/N_z will decrease with increasing Richardson number. We see that the lowest values of N_z and K_z are for the James and Vellar Estuaries. This fact is probably due to a combination of a decreased scale, or depth, effect and an increased stratification, or

TABLE 1.1 Observed Values

Location	$s_b - s_s$ ($^o/_{oo}$)	$v_s - v_0$ (cm sec^{-1})	$v_b - v_0$ (cm sec^{-1})	v_0 (cm sec^{-1})	$\partial s/\partial x$ ($^o/_{oo}$/km)	h (m)
Long Island Sound	0.8	22	−5		0.035	31
Mersey Estuary	0.9	12	−9		0.238	20.4
Southampton Estuary	1.46	8	−6		0.222	12
James Estuary	3.0	12	−10		0.587	7
Columbia River Estuary	15.0	29	−27	43	1.411	10.6
Vellar Estuary	23.5	9	−4		2.014	2.7

TABLE 1.2 Computed Values

Location	N_z (cm^2 sec^{-1})	K_z (cm^2 sec^{-1})	k $\times 10^{-3}$	i (cm/km)	K_z/N_z
Long Island Sound	8.1	9.5	11	0.03	1.2
Mersey Estuary	41	25	4.5	0.16	0.6
Southampton Estuary	12	3.3	3.3	0.12	0.3
James Estuary	4.4	2.2	0.8	0.14	0.5
Columbia River Estuary	17	8	0.7	0.54	0.5
Vellar Estuary	0.9	0.1	6.0	0.17	0.1

Richardson number, effect, particularly for the Vellar Estuary. The values for the Columbia River Estuary are high; here, however, it must be remembered that there is a strong tidal action, which in itself would indicate higher N_z and K_z values and would also have the effect of offsetting the stratification effect in the determination of the Richardson number. The Mersey Estuary shows the highest N_z and K_z values, and there is a strong tidal action here. The ratio K_z/N_z generally decreases with increasing stratification, which is indicative in part of an increasing Richardson number, with the exception of the Columbia River Estuary, for which anticipated effects have been described above.

The values of the longitudinal surface slope, i, are much the same for four of the estuaries, the Mersey, Southampton, James, and Vellar. The value of i is substantially higher for the Columbia River Estuary, related to the higher river runoff there; and it is substantially lower for Long Island Sound, related to the considerably broader and deeper area there and consequently to the lower value of v_0. Both of these effects are indeed what one might have anticipated. The values of the bottom frictional coefficient fall within the range normally found in other investigations. A word of caution should be said with regard to the higher values of k as the computations for such values become somewhat indeterminate for small values of v_b.

LONGITUDINAL DISPERSION

Considering next *longitudinal dispersion* in an estuary, which is important in the consideration of the ultimate dispersal of a pollutant, there are at least two important physical oceanographic effects that have to be considered under steady-state conditions. One is the *tidal diffusion* contribution to the total longitudinal dispersion flux and the other is the *net circulation, or velocity shear*, contribution. In terms of salt fluxes, the tidal diffusion salt flux is the integral over a tidal cycle of the product of the depth-averaged tidal velocity and tidal salinity variations. If, for example, the tidal velocity and tidal salinity variations were $\pi/2$ out of phase, this contribution would be zero. Correspondingly, the net circulation salt flux is the integral with respect to depth of the product of the net circulation velocity and the net salinity deviation from their depth-averaged means. For example, in a stratified estuary, the net circulation velocity, $v_{x1} = v_x - v_0$, will be positive and the salinity deviation, $s_1 = s - s_0$, will be negative in the upper portion of the water column and v_{x1} negative and s_1 positive in the lower portion of the water column, producing a finite value for this flux. In general, then, we anticipate that for well-mixed conditions the tidal diffusion flux contribution will be dominant and that for stratified conditions there will be a contribution from both fluxes to the total longitudinal dispersion flux with the net circulation contribution becoming dominant for estuaries in which there is little tidal action.

The basic formulations for the description of these two effects in terms of the observed salinity variations in an estuary has been given by Bowden (1962, 1963). Simplifying and assuming no lateral contributions, we then have for the mean tidal diffusion salt flux per unit breadth, S_t, from Officer (1976),

$$S_t = -\frac{1}{2} \, h \, BC \sin \delta, \tag{1.12}$$

and for the corresponding tidal diffusion coefficient, K_{xt},

$$K_{xt} = \frac{\sin \delta}{4\pi} \, C^2 T, \tag{1.13}$$

where B and C are the amplitudes of the tidal salinity and velocity variations, respectively, expressed in sinusoidal form, δ is the phase angle between the two variations, and T is the period of the semidiurnal or diurnal tide. Taking a velocity function of the form of the first term of Eq. (1.1) with the correspondingly derived salinity variation, we have for the net circulation salt flux per unit breadth, S_c,

$$S_c = -0.030 \, \frac{v_s^2 h^3}{K_z} \, \frac{\partial s}{\partial x}, \tag{1.14}$$

and for the corresponding net circulation diffusion coefficient, K_{xc},

$$K_{xc} = 0.030 \, \frac{v_s^2 h^2}{K_z}. \tag{1.15}$$

Other circulation velocity functions would give results of a similar form. The effective longitudinal dispersion coefficient, K_x, is simply the sum of Eqs. (1.13) and (1.15), or

$$K_x = K_{xt} + K_{xc}. \tag{1.16}$$

This coefficient can also be expressed in terms of simple physical oceanographic observable quantities, through the cross-sectional averaged mass and salt continuity equations, as

$$K_x = \frac{Rs_0}{A(\partial s/\partial x)}, \tag{1.17}$$

where s_0 is the cross-sectional and tidal-averaged value of the salinity.

The relation (1.17) has received a great deal of application. It provides a simple and direct expression for the determination of the longitudinal dispersion coefficient, K_x, from readily observable quantities. For most of the cases to which it has been applied, the vertical salinity distribution could be described as well mixed, so that the determined values of K_x also represent the tidal diffusion coefficient, K_{xt}. In general, then, we might expect from Eq. (1.17) that K_x should increase with river runoff for a given estuary, depending also on the longitudinal salinity gradient variation, and from Eq. (1.13) that K_x, or K_{xt}, should decrease in an upestuary direction as C decreases, depending also on the estuary geometry variations as well as on other items. These two dependencies do seem in general to hold true.

We should like to cover here a few examples of such determinations without at the same time trying to be exhaustive. Stommel (1953) applied the relation (1.17), which was derived by him, to the Severn Estuary. Bowden (1963) recalculated the values for K_x including the freshwater contributions to the Severn Estuary from its tributary rivers. The values ranged from 0.6×10^6 to 1.2×10^6 cm^2 sec^{-1}, being, contrary to the above generalization, higher up the estuary than at its mouth.

Burt and Marriage (1957) made similar calculations for the Yaquina Estuary. They found that for a high river discharge condition of 17 m^3 sec^{-1}, K_x increased in a regular manner from 0.6×10^6 to 8.5×10^6 cm^2 sec^{-1} over a longitudinal distance of 30 km from the estuary head to its mouth, and for a low river discharge condition K_x increased again in a regular manner from 0.2×10^6 to 1.0×10^6 cm^2 sec^{-1} over the same distance.

Hughes (1958) made calculations for K_x for the Mersey Estuary. He obtained an average K_x value of 1.6×10^6 cm^2 sec^{-1} for a low river discharge condition of 26 m^3 sec^{-1} and an average K_x value of 3.6×10^6 cm^2 sec^{-1} for a high river discharge condition of 103 m^3 sec^{-1}.

Bowden (1963) calculated a few selected values of the longitudinal dispersion coefficient for the Thames Estuary, obtaining 0.5×10^6 cm^2 sec^{-1} at 16 km and 0.8×10^6 cm^2 sec^{-1} at 40 km below London Bridge during low river flow conditions and 3.4×10^6 cm^2 sec^{-1} at 48 km below London Bridge during high river flow conditions.

Glenne and Selleck (1969) made an extensive study of the variations of the longitudinal dispersion coefficient as determined from Eq. (1.17) in San Francisco Bay. In the southern arm of the bay, which is well mixed and has a relatively small advective flow, the values of K_x are smaller, around 0.5×10^6 cm^2 sec^{-1}, than in the northern arm, around 2.0×10^6 cm^2 sec^{-1}. Further, the values increase with the net advective flow in the northern arm and in general increase in a downestuary direction toward the Golden Gate in both arms. Determinations of K_x were also made using silica, rather than the freshwater fraction, as a tracer and assuming that all the silica was introduced by the river flows and that it was a conservative quantity during its passage through the system. The determined K_x values show much the same distribution as those given from the freshwater fraction calculations with a somewhat larger amount of scatter.

West and Williams (1972) also made an extensive study of the K_x variations for the Tay Estuary. At the mouth of the estuary, the determined K_x values show a marked dependence on the river runoff condition varying from a value of around 1.3×10^6 cm^2 sec^{-1} at a river flow of 50 m^3 sec^{-1} to a value of around 7.0×10^6 cm^2 sec^{-1} at a river flow of 300 m^3 sec^{-1}. At the upestuary stations, the determined values were relatively insensitive to river flow, passing through a slight maximum with average values of around 1.0×10^6 cm^2 sec^{-1}.

Calculations of the longitudinal dispersion coefficient can also be made as an adjunct to finite difference, numerical model computations for dye, or dissolved oxygen distributions, using steady-state continuity relations. From dye-diffusion measurements and computations Hetling and O'Connell (1966) obtained K_x values of around 0.1×10^6 to 0.2×10^6 cm^2 sec^{-1} for the Potomac Estuary in the vicinity of Washington, D.C. From salinity calculations using the relation (1.17), they obtained K_x values increasing from this vicinity in a regular manner downestuary to around 0.6×10^6 cm^2 sec^{-1}.

From dissolved oxygen measurements and calculations, Thomann (1972) estimated K_x values increasing from 1.2×10^6 cm^2 sec^{-1} to 2.1×10^6 cm^2 sec^{-1} over a 135-km stretch of the Delaware Estuary.

There have been considerably fewer determinations of the tidal diffusion salt flux, S_t, for well-mixed or stratified estuaries and of the circulation salt flux, S_c, for stratified estuaries or of their corresponding diffusion coefficients, K_{xt} and K_{xc}.

Preddy (1954) examined the tidal mixing in the Thames Estuary. He assumed that the tidal excursion was constant along the length of the estuary, which appears to be true for the Thames but certainly is not true for estuaries in general. From his analysis, we can obtain values for sin δ, which average around sin $\delta = 0.05$, or $\delta = 3°$.

Hetling and O'Connell (1965) compared K_x values determined from Eq. (1.17) to various empirical formulas involving the tidal variables. As the Potomac is a well-mixed estuary, K_x should be equivalent to K_{xt}. From their data and analysis we can relate this to a sin δ value of around 0.20, or $\delta = 12°$.

Bowden (1960) and Bowden and Sharaf El Din (1966) calculated directly the salt fluxes S_t and S_c for the Mersey Estuary. The determined values of the S_t and S_c fluxes showed considerable variation from one measurement period to the next with the S_t values in general being larger than the S_c values. In some cases, the S_t values were of opposite sign to the S_c values, which is unusual and would indicate that on the average the water flowing out on the ebb tide had a higher salinity than the water flowing in on the flood tide. These differences are attributed to probable, but unmeasured, lateral variations.

A convenient index of the relative importance of these two contributions is the fraction, ν, which is defined as the ratio of the tidal diffusion, and possible lateral circulation, fluxes to the total longitudinal dispersion flux. Hansen and Rattray (1965) determined values for this fraction for the James Estuary and for the Columbia River Estuary, obtaining, respectively, $\nu = 0.1$ and $\nu = 0.8$–0.9. In other words, they found that for the James Estuary the circulation flux contribution was dominant and that for the Columbia River Estuary the tidal diffusion and lateral circulation fluxes were dominant. Bowden and Gilligan (1971) determined values for this fraction for the Mersey Estuary, obtaining, respectively, for four stations starting at the estuary mouth and proceeding in an upestuary direction, $\nu = 0.85$, 0.51, 0.30, and 0.62. We see, then, for the Mersey Estuary that the tidal diffusion flux is dominant at the estuary mouth and that the tidal diffusion and circulation contributions are of comparable magnitude at the other three upestuary stations.

FURTHER INVESTIGATIONS

In many respects our understanding of circulation, mixing, and related effects in estuaries is in its infancy. We have some reasonable understanding of the gross effects, but beyond that we have only a few glimpses. We do not have the same degree of understanding of the hydrodynamics of estuaries as we do of the hydrodynamics of atmospheres and oceans. Indeed, the estuarine environment is more complex than either an atmosphere or an ocean, and there are more variables that have to be considered in either theoretical derivations or field investigations. The phenomena that are observed in estuaries are not so susceptible to isolation and separate, controlled investigation.

Nevertheless, estuaries have been important in our industrial development and now, unfortunately, are important in pollution problems related to such development. We need to have a much better understanding of the scientific and engineering aspects of estuarine hydrodynamics to be able to make intelligent environmental decisions.

With regard to well-mixed and stratified estuaries, the following are some of the subjects that appear to warrant attention. Each has been the subject of investigation, and the results that have been obtained indicate that they are subjects for continued investigation. There is, of course, a certain amount of overlap and interdependency among them.

1. *Lateral Effects.* In those few estuaries where there are sufficiently detailed measurements, the lateral effects are often found to be as important as the longitudinal effects. This includes not only the lateral contributions to the longitudinal motions, such as the lateral velocity shear contribution to the longitudinal flux, but also the transverse motions themselves.

2. *Spatial and Temporal Variations.* We have a reasonable understanding of the circulation and mixing effects as averaged over a tidal cycle but not nearly so clear an understanding of the details of circulation and mixing within a tidal cycle. Geometric effects, such as lateral constrictions and shoaling, are of obvious importance but have not been studied to any great degree.

3. *Near-Source Effects.* Here we are referring in part to near-source diffusion effects within the dimensions of the tidal excursion and period where the simple mixing equations with constant coefficients do not apply. It is a scale-dependent phenomenon, both spatially and temporally; and even the crudest of dye-diffusion experiments indicate its importance and complexity.

4. *Turbulence.* A better understanding of the turbulent diffusion processes for both momentum and material transport are needed, including that related to vertical stability. The tidal and other variations of the turbulence-generated stresses within the water column and near the bottom need continued investigation.

5. *Salinity Gradient Zone.* The salinity gradient zone is the region in the upper reaches of a number of estuaries where the flow changes from that of an essentially riverine character to that of an essentially estuarine character. It is associated with the suspended sediment turbidity maximum, the accumulation zone of bottom-sediment deposition, and phytoplankton blooms, where such blooms are controlled by physical oceanographic conditions. The hydrodynamics of the salinity gradient zone are not well understood; it would appear that such factors as water particle residence time, vertical circulations, and inertial effects are considerably different here from elsewhere in the estuary or its associated river.

In addition to these strictly physical oceanographic studies there is the category of interdisciplinary processes in which circulation and mixing are important, such as suspended sediment and material transport and distribution, nutrient cycles, and larval distributions.

REFERENCES

Bowden, K. F. (1960). Circulation and mixing in the Mersey Estuary, *Proc. Internat. Assoc. Sci. Hydrol. 51*, 352.

Bowden, K. F. (1962). The mixing processes in a tidal estuary, Proceedings of the First International Conference on Water Pollution Research, in *Advances in Water Pollution Research,* Vol. 3, Pergamon Press, London, pp. 329–346.

Bowden, K. F. (1963). The mixing processes in a tidal estuary, *Internat. J. Air Water Pollut. 7*, 343.

Bowden, K. F., and R. M. Gilligan (1971). Characteristic features of estuarine circulation as represented in the Mersey Estuary, *Limnol. Oceanog. 16*, 490.

Bowden, K. F., and S. H. Sharaf El Din (1966). Circulation, salinity and river discharge in the Mersey Estuary, *Geophys. J. R. Astron. Soc. 10*, 383.

Burt, W. V., and L. D. Marriage (1957). Computation of sewage in the Yaquina River Estuary, *Sewage Indust. Wastes 29*, 1385.

Cameron, W. M., and D. W. Pritchard (1965). Estuaries, in *The Sea,* Vol. II, John Wiley and Sons, New York, pp. 306–324.

Dyer, K. R. (1973). *Estuaries: A Physical Introduction,* John Wiley and Sons, London.

Dyer, K. R., and K. Ramamoothy (1969). Salinity and water circulation in the Vellar Estuary, *Limnol. Oceanog. 14*, 4.

Glenne, B., and R. E. Selleck (1969). Longitudinal estuarine diffusion in San Francisco Bay, California, *Water Res. 3*, 1.

Hansen, D. V. (1965). Currents and mixing in the Columbia River Estuary, *Marine Technology Society and American Society of Limnology and Oceanography, Joint Conference on Ocean Science and Ocean Engineering,* pp. 943–955.

Hansen, D. V., and M. Rattray (1965). Gravitational circulation in straits and estuaries, *J. Marine Res. 23*, 104.

Hetling, L. J., and R. L. O'Connell (1965). Estimating diffusion characteristics of tidal waters, *Water and Sewage Works 110*, 378.

Hetling, L. J., and R. L. O'Connell (1966). A study of tidal dispersion in the Potomac River, *Water Resources Res. 2*, 825.

Hughes, P. (1958). Tidal mixing in the narrows of the Mersey Estuary, *Geophys. J. R. Astron. Soc. 1*, 271.

O'Brien, M. P. (1952). Salinity currents in estuaries, *Trans. Am. Geophys. Union 33,* 520.

Officer, C. B. (1976). *Physical Oceanography of Estuaries and Associated Coastal Waters,* John Wiley and Sons, New York.

Preddy, W. S. (1954). The mixing and movement of water in the estuary of the Thames, *J. Marine Biol. Assoc. United Kingdom 33,* 645.

Pritchard, D. W. (1952a). Estuarine hydrography, *Advan. Geophys. 1,* 243.

Pritchard, D. W. (1952b). Salinity distribution and circulation in the Chesapeake Bay estuarine system, *J. Marine Res. 11,* 106.

Pritchard, D. W. (1954). A study of the salt balance in a coastal plain estuary, *J. Marine Res. 13,* 133.

Pritchard, D. W. (1967). Observations of circulation in coastal plain estuaries, in *Estuaries,* Publ. No. 83, American Association for the Advancement of Science, Washington, D.C., pp. 37–44.

Rattray, M., and D. V. Hansen (1962). A similarity solution for circulation in an estuary, *J. Marine Res. 20,* 121.

Riley, G. A. (1952). Hydrography of the Long Island and Block Island Sounds, *Bull. Bingham Oceanog. Collection 13,* 5.

Riley, G. A. (1955a). Oceanography of Long Island Sound, I, Introduction, *Bull. Bingham Oceanog. Collection, 15,* 9.

Riley, G. A. (1955b). Oceanography of Long Island Sound, II, Physical oceanography, *Bull. Bingham Oceanog. Collection 15,* 15.

Stommel, H. (1953). Computation of pollution in a vertically mixed estuary, *Sewage Indust. Wastes 25,* 1065.

Thomann, R. V. (1972). *Systems Analysis and Water Quality Management,* McGraw-Hill Book Co., New York.

West, J. R., and D. J. A. Williams (1972). An evaluation of mixing in the Tay Estuary, *American Society of Civil Engineers, Proceedings of the Thirteenth Conference on Coastal Engineering,* pp. 2153–2169.

Lateral Circulation Effects in Estuaries

2

KEITH R. DYER
Institute of Oceanographic Sciences, Taunton, England

INTRODUCTION

Mathematical models of estuarine circulation have been applied to a variety of problems mainly concerned with predicting the dispersal of pollutants, both during a tidal cycle and over longer periods. Most models necessarily make an assumption about the relative importances of the vertical and lateral circulation effects.

In one-dimensional models, the estuary is considered to be sectionally homogeneous and only the longitudinal effects are included. The seaward advection of salt driven by the river discharge is balanced by a landward dispersion that can be characterized by a dispersion coefficient multiplied by the longitudinal salinity gradient. In the rather more realistic two-dimensional models, the estuary is either taken to have a lateral salinity variation and to be vertically homogeneous or to be laterally homogeneous with vertical salinity variations. The former model assumes that the lateral circulation is dominant; the latter, the vertical. Only in three-dimensional models can both lateral and vertical effects be considered.

There is little information in the literature regarding the importance of lateral variations of salinity and of velocity in the dynamic and salt balances. Such information as there is suggests that in many instances the lateral and vertical circulations are of a similar order of importance (Dyer, 1974) and that three-dimensional models may generally be required.

SECONDARY FLOWS

The vertical gravitational circulation in estuaries is caused by the mixing between the dense salty seawater and the fresher less dense river water. This fact, coupled with the requirement that the fresher water is driven by the river discharge toward the sea, gives a vertical movement of water and an inflow of saltwater near the bed, which creates a vertical salinity gradient. The strength of the circulation will depend on the efficiency of the mixing process and on the estuary breadth and depth. These factors have been examined theoretically by Hansen and Rattray (1965) for laterally homogeneous conditions in a rectilinear estuary.

In practice, however, the flows involved in these circulations will not be evenly distributed across the estuary cross section, as the water tends to flow in a spiral fashion. The transverse and vertical components of the spiral flow create what are known as secondary flows in the plane of the cross section. These flows arise from the combination of several interrelated effects. First, because the cross section of the estuary is not rectangular, the vertical exchange between the saltier water in the deeper part and the fresher water is unevenly distributed. Second, the cross-sectional form of the estuary is not consistent longitudinally. This not only creates a speeding up of the flow as the shallower water is encountered but also causes a deflection of the flow toward the deeper parts of the cross section. Third, flow around bends in the estuary causes the thread of the maximum current to swing toward the outside of the bend. Bends also have changes in cross-sectional form associated with them, deep scour holes being present on the outside of the bend.

The actual components of the secondary circulation are difficult to measure. The lateral components can be measured with conventional current meters; but as they are components of large longitudinal flows, relatively large errors are produced by small errors in measurement of the current direction. Averaging over a tidal cycle does tend to randomize these errors, however. The vertical components can only be calculated at the moment from continuity principles using tidally averaged data. This involves measurements of the longitudinal and lateral components at a grid of stations whose spacing is sufficiently far apart for the velocity differences between stations to be significantly larger than the measurement errors, yet sufficiently close that the assumption of a linear gradient in the velocities is justified. The vertical velocities calculated by this means are thus an average not only over the tidal cycle but also over a considerable area. The magnitude of the maximum mean longitudinal, lateral, and vertical velocities are often in the ratio of about 1000:100:5. Despite the errors, some general principles appear, and three basic secondary circulation patterns can be recognized.

Firstly, in rivers it has long been known that the convergence of the longitudinal current with the outside bank on a bend causes a downward flow into the deep scour hole and a rise in the water surface—a super elevation—on the outside of the bend (Chow, 1959). On the shallower inside of the bend, there is a divergence of water from the bank and an upward flow. Thus on the left-hand bend there is a clockwise motion. As the meander bends alternate, the secondary flow has to change its sense before the right-hand bend downstream. This is effected in the straight channel section between bends where there is often a longitudinal central bar separating the two cells. This circulation is illustrated in Figure 2.1. Similar flow patterns exist even in straight channels (Einstein and Shen, 1964) and appear to hold also in the surface layer of salt-wedge estuaries. In this case, in spite of a degree of tidal oscillation, the saltwater does not mix significantly with the freshwater above. Within the salt wedge there is a slight landward inflow, which has lateral components imposed on it by changes in the channel position, but it is not strongly coupled to the surface circulation. However,

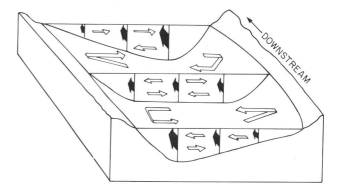

FIGURE 2.1 Diagrammatic representation of mean current flow in a river or a salt-wedge estuary. Not to scale.

the cross-channel slope of the halocline and the zone of zero net motion marking the upper surface of the salt wedge will respond to the dynamics of the flow above. It appears that the salt wedge is higher on the inside of the bend than on the outside.

Secondly, in partially mixed estuaries there is a significant exchange between the surface and bottom layers. This involves an upward transfer of salt and an upward flow of water the magnitude of which is governed by the overall dynamics of the estuary. To compensate for the vertical exchanges, there is an appreciable mean landward flow within the salt intrusion whose magnitude diminishes toward the head of the estuary (Pritchard, 1952; Dyer, 1973). In the shallower parts of the cross section, the full salinity profiles are not developed, only the top parts being present. The surface is fresher than over the deep channel, and, because of the increased friction caused by the shallower depth, the profile is generally better mixed and the bottom salinity is less than that at an equivalent depth in the main channel (Figure 2.2). The water over the deep channel is maintained more salty by the vertical exchange of salt with the lower layer. Consequently, the secondary current is now upward over the deep water and downward over the

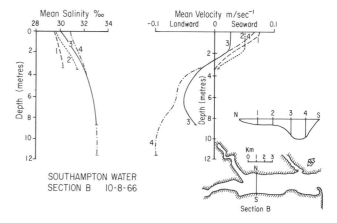

FIGURE 2.2 Mean salinity and velocity profiles for four stations across Southampton water.

shallow. Thus, on a left-hand bend the secondary circulation is anticlockwise (Figure 2.3), although it is less intense than in salt-wedge estuaries. The halocline and the depth of no motion now slope downward toward the inside of the bend. The effects are the average of those occurring during the tidal cycle. Field information is generally not good enough for the instantaneous secondary circulation to be calculated. The change in the sense of the secondary circulation with the change from salt wedge to partially mixed type of estuary has been documented for the Vellar Estuary during a decrease in river flow (Dyer and Ramamoorthy, 1969). It appears that the lateral circulation will contribute more to the salt balance in the partially mixed than in the salt-wedge estuaries, even though the mean velocities are smaller.

Thirdly, in wide, shallow estuaries where the tidal amplitude is large the mixing will be sufficiently intense to create vertically homogeneous conditions. The tidal excursion is large, and the maximum flood current takes a different position in the cross section than the maximum ebb. In the deeper channel, the ebb currents are normally dominant and give a strong downstream residual current over the tidal cycle. In the shallower parts, the flood currents are dominant. Consequently, a horizontal circulation is present (Figure 2.4). The Gironde Estuary is an example of this, on spring tides and at low river discharge (Allen, 1973). The mean flow is seaward in the deeper channel on the left-hand side (Figure 2.5) and causes a large seaward salt transport. The mean flow is also seaward over the rest of the channel, but because of the progressive nature of the tidal wave, slack water occurring almost an hour after high and low water, the mean salt transport in those parts is actually landward. The maintenance of fresher water on the right-hand side results from strong cross-channel exchanges at certain stages in the tidal cycle.

LATERAL DYNAMIC BALANCE

The secondary flow patterns outlined above produced cross-channel accelerations that must be in balance with the driving forces. This balance has been investigated by Pritchard (1956) and Dyer (1973), and it is of interest to examine

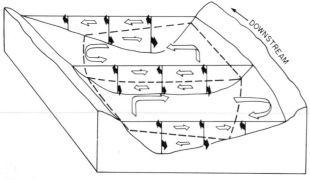

FIGURE 2.3 Diagrammatic representation of mean current flow in a partially mixed estuary. Not to scale.

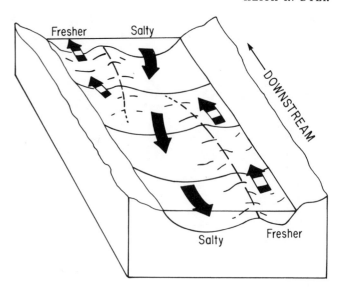

FIGURE 2.4 Diagrammatic representation of mean salt transport in a wide vertically homogeneous (well-mixed) estuary.

the relative importances of the various items in the equation of conservation of momentum.

Considering a right-handed coordinate system located at the water surface with the tidal mean velocities $\bar{u}$, $\bar{v}$, and $\bar{w}$ in the longitudinal x, lateral y, and vertical z directions, respectively, the lateral equation of motion is

$$\frac{\partial \bar{v}}{\partial t} + \bar{u}\frac{\partial \bar{v}}{\partial x} + \bar{v}\frac{\partial \bar{v}}{\partial y} + \bar{w}\frac{\partial \bar{v}}{\partial z} = -\frac{1}{\rho}\frac{\partial p}{\partial y} - g\frac{\partial \zeta}{\partial y}$$

$$\quad 1 \qquad 2 \qquad 3 \qquad 4 \qquad\qquad 5 \qquad\quad 6$$

$$-f\bar{u} + \frac{\bar{u}^2 + \bar{U}^2}{-R} + \frac{1}{\rho}\frac{\partial F}{\partial z}$$

$$\qquad 7 \qquad\quad 8 \qquad\quad 9$$

$$-\frac{\partial \overline{(u'v')}}{\partial x} - \frac{\partial \overline{(v'v')}}{\partial y} - \frac{\partial \overline{(w'v')}}{\partial z}. \qquad (2.1)$$

$$\qquad 10 \qquad\quad 11 \qquad\quad 12$$

The overbars denote averaging over a tidal cycle. Term 1 is the time change in the lateral velocity, normally taken to be zero. Terms 2 to 4 are the field acceleration terms associated with the secondary circulation. Term 5 is the horizontal pressure force created by differences in the water density, ρ being the density and p the pressure. Term 6 is the horizontal pressure force resulting from differences in the elevation ζ of the water surface across the estuary. This term usually cannot be measured directly. Term 7 is the Coriolis force acting toward the right of the mean longitudinal current in the northern hemisphere, f being the Coriolis parameter ($0.729 \times 10^{-4} \sin\phi$, where ϕ is the latitude). Term 8 is the centrifugal force arising from curvature of the streamlines

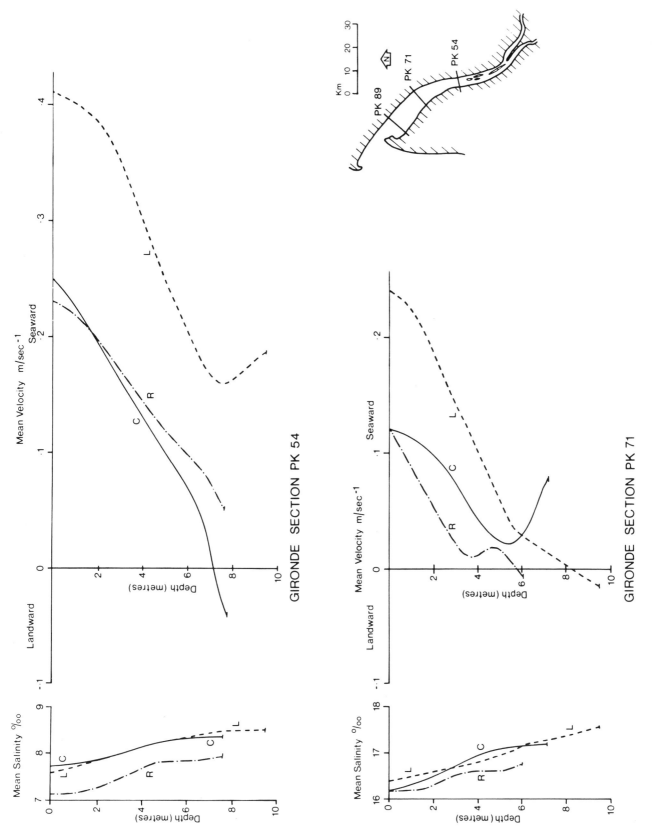

FIGURE 2.5 Mean salinity and velocity profiles for three stations on two cross sections across the Gironde Estuary.

of radius R. U is the tidal oscillation velocity. Term 9 is the lateral frictional force created by wind stress on the sea surface. This term can sometimes be written as $(F_s - F_B)/\rho h$, where F_s and F_B are the surface and seabed shear stresses and h is the water depth. Terms 10 to 12 are the turbulent Reynolds stresses, u', v', and w' being the turbulent deviations from the sum of the mean and tidal oscillation velocities.

In rivers, the balance is normally assumed to be between the surface water slope and the centrifugal force, although more accurate results are obtained by including a circulation constant that represents the contribution of the secondary circulation (Chow, 1959). In estuaries, the oscillating flows and the presence of internal density gradients add complicating factors.

The relative magnitudes of the various terms have been calculated for the Vellar Estuary. The results are shown in Table 2.1 for the surface and 0.5-m depth between two stations during a river flow period when the estuary was partially mixed. The time change in the lateral velocity and the wind stress are assumed to be zero, which are considered to be assumptions in this case. The surface-water slope and the radius of curvature R are initially unknown. The Reynolds stress terms are assumed to be negligible. Terms 4 and 5 are obviously zero at the surface. Let us assume that the radius of curvature is constant between the two depths. Then, by taking the differences between the values at the two depths, the surface-slope term 6 disappears and the radius of curvature can be calculated as 2.38 km. This is consistent with the estuary outline. Using this value of R the centrifugal force, term 8, and thence the surface-water slope, can be calculated at both levels. A consistent value of surface-water slope of 7.38×10^{-5} is obtained.

This analysis shows that the dominant terms in the lateral dynamic balance are the water slope, the internal density structure, and the centrifugal force. Coriolis force appears to be of secondary importance. The cross-flow terms represent only a small imbalance between the forces. Similar analysis of results from Southampton water give the same conclusions. For the Vellar Estuary, when of a salt-wedge type, the lateral balance in the surface layer is, like a river, mostly between the water slope and centrifugal force. For larger estuaries, however, the Coriolis force is likely to become relatively more important as the centrifugal force decreases.

WIND EFFECTS

Of all the terms in Eq. (2.1), however, wind stress is potentially the most significant. A surface wind stress of 0.1 dyne cm^{-2}, equivalent to a 2.5 m sec^{-1} wind (Force 2), gives a value of 250×10^{-6} m sec^{-2} for term 9 in a water depth of 4 m. Because the wind is likely to vary with a time scale similar to the tidal period, its effect on the lateral balance will be large, especially in shallow estuaries. When the wind starts blowing, the whole mass of water will tend to be driven across the estuary. At this time, F_B may be significant and would reduce the effective value of term 9 in Eq. (1). Also, the time change term in the lateral velocity would be large. The water movement would cause a setup in the water surface, eventually leading to a large value for term 6 to oppose the wind stress. This process would occur fairly quickly. However, beneath the wind a water circulation would be created, giving downwelling on the downwind side and upwelling on the upwind side of the estuary. This secondary circulation would lead to negative values of F_B and an increase in the frictional term, which would be balanced by adjustment of the internal density distribution. The vertical mixing enhanced by the wind will decrease the stratification and alter the density structure. A decrease in the wind would cause a rapid readjustment of the water slope, with a more delayed response of the circulation and density distribution.

The effect of the wind is likely to vary considerably along the estuary, with sheltering and fetch. Much of the inconsistency in the results of calculations of salt transport and mean circulation patterns in estuaries derived from short time series of data is probably the result of wind effects and also of barometric pressure, which can increase or decrease the total water volume of an estuary by a significant amount.

DISPERSION

Studies of pollutant dispersal are aimed at predicting in the near field the manner in which the pollutant plume spreads, and in the far field the concentration variation as a function of distance from the source. The dispersion is carried out by tidal diffusion and by velocity shear. The latter contribution is caused by pollutant trapped in slower

TABLE 2.1 Lateral Dynamic Balance between Two Stations in the Vellar Estuary

	$\bar{u}\dfrac{\partial \bar{v}}{\partial x}$	$\bar{v}\dfrac{\partial \bar{v}}{\partial y}$	$\bar{w}\dfrac{\partial \bar{v}}{\partial z}$	$\overline{\dfrac{1}{\rho}\dfrac{\partial p}{\partial y}}$	$f\bar{u}$	$\bar{u}^2 + \overline{U^2}$	$\dfrac{\bar{u}^2 + \overline{U^2}}{-2380}$	$g\overline{\dfrac{\partial \zeta}{\partial y}}$
Term	2	3	4	5	7		8	6
Depth, m	10^{-6} m sec^{-2}					$m^2 sec^{-2}$	10^{-6} m sec^{-2}	
0	−0.25	0.02	0	0	1.10	0.0155	−6.51	7.38
0.5	0.51	−1.44	0.56	5.92	0.87	0.0023	−0.97	7.39

running water near the bed or sides of the estuary gradually mixing back into the main body of the flow. This causes the pollutant to spread out across and along the estuary. Calculation of the velocity shear effect from first principles has been attempted by Fischer (1972) and Okubo (1973). The tidal diffusion occurs because of the tidally oscillating flow and the facts that ebb and flood velocities are usually unequal and there are phase shifts between the velocity, concentration, and cross-sectional area variations (Holley *et al.*, 1970). Dispersion is also increased by bends and is related to aspect ratio (width/depth).

Dispersion coefficients that can be used for prediction can be calculated from the observed salinities and velocities measured on a cross section by a process outlined by Fischer (1972) and extended to include the tidal diffusion terms by Dyer (1974). This separates the measured values into cross-sectional mean terms and a series of terms resulting from the deviations of the mean and oscillatory values at any depth and position from the cross-sectional mean values. The vertical and transverse effects are separated so that their individual contributions can be compared. Omitting a number of terms that are relatively unimportant, this gives a mean flux of salt of

$$\bar{F} = \bar{A}\bar{u}_A\bar{s}_A + \overline{AU_A\bar{s}_A} + \overline{AS_A\bar{u}_A} + \overline{AU_AS_A} + \overline{A(\bar{u}_t\bar{s}_t)_A}$$

$$\quad 1 \qquad\quad 2 \qquad\quad 3 \qquad\quad 4 \qquad\quad 5$$

$$+ \ \bar{A}(\bar{U}_v\bar{S}_v)_A + \bar{A}(\overline{U_tS_t}) + \bar{A}(\overline{U_vS_v})_A$$

$$\qquad 6 \qquad\qquad 7 \qquad\qquad 8$$

(2.2)

where overbars denote averaging over a tidal cycle and a subscript A denotes averaging over the cross section. Subscripts t and v denote the transverse and vertical deviations from the cross-sectional means. U, S, and A are the tidally varying velocity, salinity, and cross-sectional area, $\bar{u}$, $\bar{s}$, and $\bar{A}$ are the tidal mean values. Term 1 is the result of the river flow. Term 2 is the compensation for the inward transport on the partially progressive tidal wave. Term 3 is the correlation between tidal variations in area and salinity. Term 4 is the correlation between the tidal oscillations in velocity and salinity. Terms 5 and 6 are the contributions of the net transverse and vertical gravitational circulations. Terms 7 and 8 are the transverse and vertical oscillatory shear. Terms 1 and 2 do not contribute to the dispersion.

Terms 3 and 4 are the tidal diffusion terms, and terms 5 to 8 are the shear effects. A convenient measure of the relative importance of the two contributions is the ratio, ν, which is

$$\nu = \frac{\text{Term 3} + \text{Term 4}}{\Sigma\,\text{Term 3 to Term 8}}. \qquad (2.3)$$

Evaluation of the magnitudes of these terms for five estuaries has shown that there are significant differences between salt-wedge, partially mixed, and well-mixed estuaries. Comparative values, together with the calculated values of ν, are shown in Table 2.2. Errors in term 1 are often considerable, but errors are usually small for the other terms (Dyer, 1974).

For the salt-wedge estuary, the mean vertical gravitational circulation is dominant over the transverse. In the partially mixed estuary, the mean transverse and vertical terms (5 and 6) are similar in magnitude, and lateral effects are important in maintaining the salt balance. The oscillatory terms are small, but in other partially mixed estuaries with higher tidal currents terms 7 and 8 and the tidal diffusion terms are considerably increased. Also, the aspect ratio of the cross section appears to be important in determining the relative magnitude of the transverse terms. In the well-mixed estuary, the oscillatory terms are of a similar magnitude to the mean terms, but by far the biggest contribution to maintaining the salt balance is the tidal diffusion arising from the phase differences between the tidal variations of salinity, velocity, and cross-sectional area. This must involve fairly widespread lateral movement of water at or near slack water, which does not mix extensively with the surrounding water. This is enhanced by the topographic separation of the estuary into three longitudinal channels.

LATERAL STRATIFICATION

The degree of lateral stratification can be thought of in terms of the stratification–circulation diagram of Hansen and Rattray (1966). Five estuaries are considered, each with at least three stations on a cross section. On three of the estuaries there are two or more sections, and on one of the estuaries there are four discharge stages. The normal classification of these estuaries is shown in Figure 2.6, in which the mean surface velocity u_s is the average value of those measured on the cross section. A lateral stratification parameter $\Delta s/s_0$ has been calculated from the maximum differences in

TABLE 2.2 Salt and Balance in Three Estuaries of Different Type (in kg sec^{-1})

	$\bar{A}\bar{u}_A\bar{s}_A$	$\overline{AU_A\bar{s}_A}$	$\overline{AS_A\bar{u}_A}$	$\bar{A}\,\overline{U_AS_A}$	$\bar{A}(\bar{u}_t\bar{s}_t)_A$	$\bar{A}(\bar{u}_v\bar{s}_v)_A$	$\bar{A}(\overline{U_tS_t})_A$	$A(\overline{U_vS_v})_A$	ν
	1	2	3	4	5	6	7	8	
Salt wedge (Vellar)	87.8	−52.3	2.1	−31.1	−6.1	−195.8	3.7	−20.0	0.10
Partially mixed (Southampton water)	−2961.0	−84.0	−1.7	2.1	−223.8	−198.0	−21.1	−12.4	0.01
Well mixed (Gironde)	−25301.3	−18718.6	1478.1	−5855.9	138.1	−545.5	−133.7	−252.9	0.83

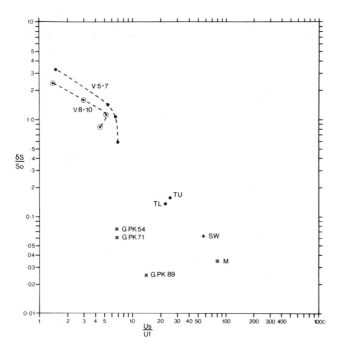

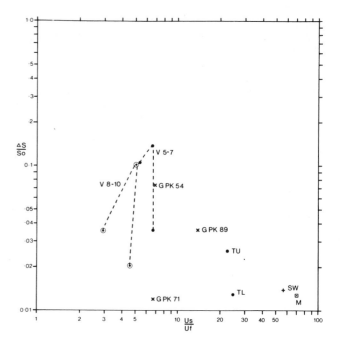

FIGURE 2.6 Classification of five estuaries on a stratification-circulation diagram. U_s is the mean surface velocity, U_f the cross-sectional mean flow associated with the river discharge, δS the difference in mean salinity between the surface and bottom, and S_o the cross-sectional mean salinity. $V5^.$-7 and $V8^.$-10 are for two sections of the Vellar at four river flow stages, G, the Gironde at the three sections shown in Figure 2.5. TL and TU are two sections on the Tees estuary, SW is Southampton water, and M the Mersey narrows.

FIGURE 2.7 Lateral stratification–circulation diagram for the five estuaries shown in Figure 2.6. ΔS is the lateral difference in mean surface salinity.

the mean surface salinities between stations on the cross section, normalized by the cross-sectional mean salinity. This gives a parameter similar to the vertical stratification parameter. The results for the Vellar Estuary are of particular interest (Figure 2.7). At highest river flow the surface layer is freshwater, but as the river flow diminishes there is a maximum relative lateral stratification. This point coincides with the reversal in the sense of the secondary circulation pattern and also with the period of minimum mixing calculated by Dyer and Ramamoorthy (1969). The lateral stratification then decreases with further decreasing river discharge.

From comparison of the results between the two sections on the Vellar and those on the Gironde it is also apparent that for any particular estuary the lateral stratification decreases for increasing aspect ratio. For most estuaries this means that the lateral stratification decreases toward the mouth. The Tees is an exception to this, which may be because of its trained banks and dredged channel. The lateral stratification is significant in all cases, however.

CONCLUSIONS

There are significant differences in lateral circulation effects between salt-wedge, partially mixed, and well-mixed estuaries. These are apparent in the secondary circulation patterns and in the relative importances of lateral terms in the dynamic balance and in the salt transport. In the five estuarine examples so far examined, lateral variations are significant and lateral homogeneity does not appear to be a widely valid assumption. Mathematical modeling, particularly of partially mixed estuaries, ought to take account of this. Also, in well-mixed estuaries the phase relationships between the tidal variations of salinity and velocity and in cross-sectional area during the tide are significant. Wind-induced changes in the lateral circulation can be expected especially in shallow estuaries.

ACKNOWLEDGMENTS

I would like to thank G. P. Allen (CNEXO, Brest) for allowing me to use the data from the Gironde and R. E. Lewis (ICI, Brixham) for permission to use those from the Tees.

REFERENCES

Allen, G. P. (1973). Etude des processes sedimentaires dans l'estuaire de la Gironde, *Mem. Inst. Geol. Bassin Aquitaine,* No. 5.

Chow, V. T. (1959). *Open-Channel Hydraulics,* McGraw-Hill Book Co., New York.

Dyer, K. R. (1973). *Estuaries: A Physical Introduction,* John Wiley and Sons, London.

Dyer, K. R. (1974). The salt balance in stratified estuaries, *Estuarine Coastal Marine Sci. 2,* 273.

Dyer, K. R., and K. Ramamoorthy (1969). Salinity and water circulation in the Vellar Estuary, *Limnol. Oceanog. 14,* 4.

Einstein, H. A., and H. W. Shen (1964). A study of meandering in straight alluvial channels, *J. Geophys. Res. 69,* 5239.

Fischer, H. B. (1972). Mass transport mechanisms in partially stratified estuaries, *J. Fluid Mech. 53,* 671.

Hansen, D. V., and M. Rattray, Jr. (1965). Gravitational circulation in estuaries, *J. Marine Res. 23,* 104.

Hansen, D. V., and M. Rattray, Jr. (1966). New dimensions in estuary classification, *Limnol. Oceanog. 11,* 319.

Holley, E. R., D. R. F. Harleman, and H. B. Fischer (1970). Dispersion in homogeneous estuary flow, *Am. Soc. Civil Eng. J. Hydraulics Div. 96,* HY8, 1691.

Okubo, A. (1973). Effect of shoreline irregularities on streamwise dispersion in estuaries and other embayments, *Netherlands J. Sea Res. 6,* 213.

Pritchard, D. W. (1952). Estuarine hydrography, *Advan. Geophys. 1,* 243.

Pritchard, D. W. (1956). The dynamic structure of a coastal plain estuary, *J. Marine Res. 15,* 33.

River Plumes
and
Estuary Fronts

3

RICHARD W. GARVINE
The University of Connecticut

INTRODUCTION

The manner in which freshwater flows down to the sea and mixes with it is one that varies greatly in nature. The extremes of the possibilities are marked, on the one hand, by a wide estuary where seawater moves inland well upstream of the mouth to meet the freshwater and, on the other hand, by a river whose lower valley is narrow, which confines its discharge so that the freshwater is forced out at its mouth to mix with salt water only offshore. This latter condition, in which a distinct plume of brackish water is produced beyond the mouth, is the principal concern of this chapter. Of course, there are many examples of estuaries to be found that fall between these limits.

What is it that determines where the mixing of freshwater and salt water will mainly occur? There are two chief parameters—the mean velocity downriver of freshwater and the root-mean-square tidal current velocity. For fixed tidal amplitude, an increase in freshwater discharge velocity will push the mixing zone downstream and ultimately into the receiving body offshore. For fixed discharge, an increase in tidal current will invigorate the mixing process between

water types and diffuse salt further upstream. The ratio of these two velocities has been introduced by Hansen and Rattray (1966) as a convenient index for classifying estuarine salinity distributions and is designated here by the symbol P. From observations of surface salinity we can examine how the upstream distance of a particular isohaline changes with P. Figure 3.1, for example, shows for the Connecticut River Estuary how the distance upstream of the mouth for a surface salinity 10 percent of that of the maximum varies with P for times of both high and low slack tidal current. As one might anticipate, as P increases from values small compared with 1, the distance decreases rapidly until, for low slack water, the river is nearly free of salt on the surface when $P \approx \frac{3}{4}$. For values above $P = 2$, even at high slack water, the river is nearly salt-free, and the 10 percent location reaches an asymptotic location offshore of the mouth. Based on these data and other observations, we may assert that a well–defined river plume will exist beyond a river mouth following ebb tide at least when $P > \frac{3}{4}$ and for all tidal stages when $P > 2$. From the definition of P, then, distinct plumes are associated with estuaries in which the average downstream speed of freshwater discharge is as

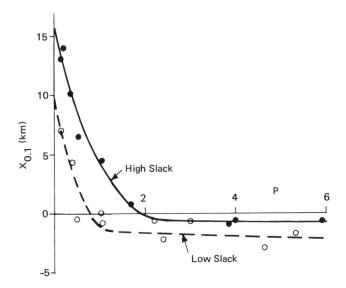

FIGURE 3.1 Distance from the mouth (positive upstream) of the point where surface salinity is 10 percent of the maximum found versus P for high and low slack tidal current in the Connecticut River Estuary. From Garvine (1975).

large or larger than the tidal current amplitude; hence, only weak upstream flow, if any at all, can be expected during flood tide. Rivers with distinct plumes under mean conditions include the Amazon, the Columbia, the three major passes of the Mississippi River delta, and the Connecticut. Other well-known rivers that do *not* generally form plumes and have P values small compared with 1 are the Hudson, the Delaware, and the St. Lawrence.

PHYSICAL PROPERTY DISTRIBUTIONS IN RIVER PLUMES

In the estuarine zone, distribution of properties is controlled mainly by the mixing of two distinct water masses, viz., freshwater from upriver above the reach of salt intrusion and adjacent coastal seawater. We may regard each water mass as a reservoir within which properties vary slowly in time and space in comparison with the variations in the estuarine zone. Offshore there will exist a body of seawater with fairly uniform salinity and temperature, which vary somewhat over a year. We may view upstream river water as a reservoir of zero salinity and of a temperature determined by the time of year related to varying temperature in the upland. In the mixing region, we find only levels of temperature and salinity that are between the extremes of these two variables. The details of the property distributions will then be set mainly by the characteristics of the turbulence field that produces the mixing. The main source of energy for the turbulence is, of course, the tidal current. Since the turbulence field seems to mix all passive physical scalar properties in the same way, once the distribution of one such property, say salinity, is known, the distributions of other such prop-

erties, such as temperature or density, may be predicted, given their values in the two reservoirs. The correlation of these properties, in other words, will be high. For the Connecticut River, for example, the correlation coefficient for salinity and temperature was found (Garvine, 1975) to be typically 0.98. Nonconservative properties, such as dissolved oxygen and pH, will not, in general, have similar distributions, but the passive physical properties do provide background conditions for them.

The horizontal extent of river plumes depends on their mean discharge level. The Amazon River has, not surprisingly, the largest identifiable plume, several millions of square kilometers in area within the Atlantic Ocean. The Columbia River plume measurably dilutes an area of about 10^5 km² in the Pacific Ocean. Rivers with smaller discharge levels (but with P large enough) form smaller plumes. That of the Connecticut River is typically 50 km² in area.

Patterns of river plume salinity distributions show three major attributes. Plumes are shallow pools of lighter water overlying more dense ambient water; there is a very high vertical gradient in density, which, in turn, inhibits vertical mixing; and, frequently, the lateral boundaries of plumes are formed by sharp fronts.

Plumes are exceedingly thin features by almost any geometric standard. For example, the typical depth of both the Connecticut River plume in Long Island Sound (Garvine, 1974a) and that of the Mississippi River at South Pass (Wright and Coleman, 1971) is 1 m, but the horizontal extent is typically 10 km. Thus, the depth-to-width ratio is about 10^{-4}, the same as for the oceans or for a sheet of paper. Figure 3.2 shows the isohalines (contours of uniform salinity) at a depth of 0.5 m for the plume of the Connecticut River at low slack and high slack water. The plume is more suggestive of a puddle than of a jet.

Unlike the oceans, whose thin figure is dictated by geomorphology, the thinness of river plumes is a result of the high static stability that derives from the high vertical gradient of density. Although the density differences are only a percent or two between the two reservoirs, they are, nevertheless, among the highest differences found in natural bodies of water. As a result of this high stability, vertical mixing between plume and ambient water is much inhibited and the interfacial zone separating the two layers of water along the horizontal is sharply confined. Figure 3.3 is a plot of isolines of uniform density (isopycnals) in a vertical section along the plume centerline of the Connecticut River in Long Island Sound. The vertical scale is greatly expanded. The density is indicated in sigma-t units, i.e., 10^3 times the difference between the water density (in g cm^{-3}) and 1. The interfacial region is typically only 0.5 m thick and ascends from about 2-m depth near the mouth to about 1 m several kilometers from the mouth.

A vertical section along the line marked "Transect A" in Figure 3.3 and normal to the plume axis is shown in Figure 3.4. Here, in addition to the sharp interface beneath the plume, we see its counterpart at the offshore plume edge. Plume water terminates abruptly, not gradually, at a front

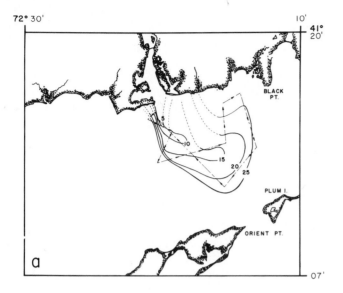

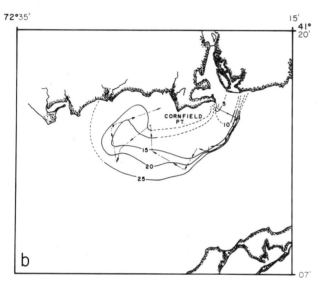

FIGURE 3.2 Lines of constant salinity in parts per million at 0.5 m depth for the Connecticut River Plume in Long Island Sound on April 21, 1972, at (a) low slack water and (b) high slack water. Dashed lines with arrow heads indicate ship track. From Garvine (1974a).

where the isolines suddenly become very steep in slope (relative to the rest of the plume) and many crop out at the surface. The term "front" has been borrowed from the atmospheric science literature; in these frontal areas, as at the edge of river plumes, are found some of the sharpest water property horizontal gradients in nature. Since the color of plume water nearly always differs substantially from that of the ambient seawater, the frontal boundary is clearly marked in photographs of the surface, such as the one shown in Figure 3.5. As is often the case, a line of foam is concentrated along with other floating debris at the surface front by convergent motion. This arrangement of properties, so near to an ideal

discontinuity, underscores the fact that river plumes need not occur merely as smooth blends offshore of upland water into seawater, as one might well have expected, but instead employ both gradual and sharp changes in their adjustment to ambient levels.

RIVER PLUME DYNAMICS

The motion field of a river plume is complex. In most cases tidal motion is either dominant or at least highly significant but serves mainly to move salt and plume water together without much altering their *relative* motion. For plume dynamics, the most important motion is that resulting from the buoyant outspreading of lighter water over heavier salt water in the shallow pool of the plume, somewhat as one would expect a pool of oil to spread over a deep layer of water or cream over milk. The leading edge of this spreading action is the frontal zone.

To gain a more detailed understanding of plume water motion we can examine the velocity field shown in Figure 3.6. Velocity vectors are drawn at grid spacings of 2 km for both the surface and a depth of 4 m, i.e., substantially below plume water. These vectors were calculated by interpolating the velocities of a large ensemble of drogues deployed in, below, and adjacent to the plume of the Connecticut River in Long Island Sound during an ebb stage during an aerial photography experiment performed by the author on May 17, 1975. The subsurface velocity field shows the nature of the basic tidal motion on the ebb toward east. Because of differences in tidal phase and water depth, the ebb current diminishes near the Connecticut shore before it subsides offshore; this explains the appearance in Figure 3.6 of stronger ebb flow further offshore in deeper water. The location of the surface front at the offshore edge of the plume is shown. Subsurface motion of salt water does not appear to have been influenced by the plume; this motion, instead, varies smoothly from the shoreline out. However, for the surface motion, the front again served as a surface of near discontinuity. Surface plume water was drawn rapidly toward the front over nearly all of the plume. Typical velocities in that

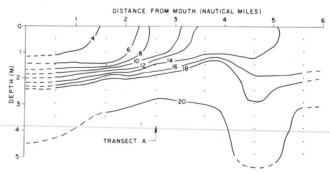

FIGURE 3.3 Isopycnals on a vertical section along the plume axis of the Connecticut River in Long Island Sound on April 13, 1973. From Garvine (1974a).

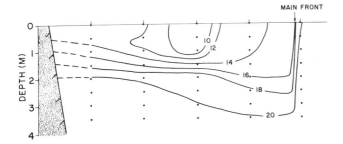

FIGURE 3.4 Isopycnals in a vertical section normal to the axis of the Connecticut River Plume marked by Transect A in Figure 3.3. From Garvine (1974a).

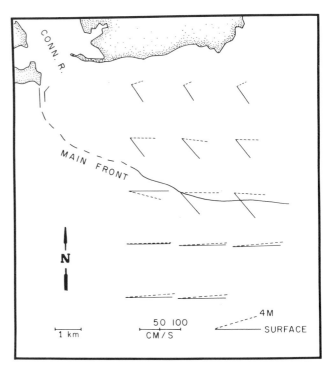

FIGURE 3.6 Horizontal velocity vectors for surface (solid lines) and 4-m depth (dashed lines) at points on a 2-km grid as computed from drogue tracks in and near the plume of the Connecticut River on May 17, 1975.

direction were about 50 cm/sec, while the front itself was found to move at about 20 cm/sec normal to itself (southerly) as the plume water spread farther laterally into Long Island Sound. This discrepancy in speed shows that surface plume water was sinking rapidly near the front to make way for yet more plume water coming behind it. Hence, a surface convergent motion was well established there. In general, the difference, both in direction and magnitude, of plume surface and subsurface (4-m-depth) water was remarkable. Beyond the front, however, the ambient salt water was moving with nearly the same speed and direction as that at depth; it was simply part of the ebbing tidal current.

The stucture of the velocity field shown in Figure 3.6 is not simple. The plume is no mere jet, nor does its fluid simply seep out and mix gradually with ambient water. Instead, a vigorous flow is produced both laterally, normal to the plume axis, and along the axis, which differs sub-

stantially from the tidal current field. Clearly, the presence of the front is fundamental to the plume dynamics, as well as to its property distribution.

ESTUARY FRONTS

Estuary fronts justify study for the sake of understanding both their own local phenomena and their strong influence on the larger scale. Awareness of their existence is still growing, but from observations to date it appears that the most common locations are along the offshore borders of river plumes and within the broader estuaries, such as the Delaware Bay. Those thus far investigated all display very large horizontal gradients in salinity, temperature, density, and color; strong convergence at their surface manifestations with a resulting accumulation there of floating material; and frequently a pronounced shear across the front of the current component parallel to the front. The river plume fronts are usually stronger in gradients and relative fluid motion, but those of the broad estuaries appear no less important to the overall motion field and property distributions on the scale of the estuary itself. This latter type has been investigated by Szekielda *et al.* (1972) in the Delaware Bay and by Ingram (1976) in the St. Lawrence River over the Ile Rouge bank. In both of these studies, tidal motion over topographic features within the estuary

FIGURE 3.5 Aerial photograph showing boundary of the Connecticut River plume on April 26, 1972. Plume surface water is lighter toned and to the right.

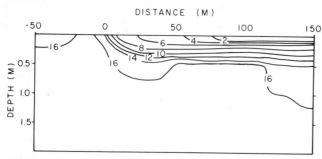

FIGURE 3.7 Density section normal to the front of the Connecticut River Plume. The origin of coordinates marks the location of the observed surface color boundary. From Garvine and Monk (1974).

appeared to be of primary importance in generating a front during each tidal cycle.

Some of the observed features of estuary fronts are reviewed here. Let us consider the distribution properties and fluid motion as seen relative to the moving coordinate system of the front itself. Figure 3.7 shows the density structure for a vertical section normal to the front, typical of a river plume front. The isopycnals reach an asymptotic depth at a distance of about 50 m from the surface location of the color boundary (which is at the origin) and have a characteristic slope of 10^{-2} in the frontal zone. Figure 3.8 shows a typical profile of the horizontal current component normal to the front at a station 30 m to the freshwater side of the color boundary. Near-surface water is moving toward the front at speeds up to 40 cm/sec, while below a depth of 2 m the motion of salt water is in the opposite direction at about 50 cm/sec. This latter speed mainly appears as a result of the spreading motion of the frontal boundary itself, whose motion we are following. Below 2 m, the vertical variation in speed is much less than that above. A strong surface convergent motion is indicated within the frontal zone. The near-surface water motion does not continue on beyond the front. Instead, the surface water is turned downward and continues with a reversal of direction along the lower portion of the light-water pool. This concept of motion in the frontal zone is consistent with the larger-scale motion of plume water displayed in Figure 3.6.

The motion in the frontal zone invites explanation. We can, in fact, explain the dynamics from fairly simple fluid-

dynamic fundamentals. The pool of lighter water in the frontal zone is in isostatic equilibrium vertically and thus floats upon the heavier ambient water. Thus, where vertical columns of lighter water within the pool penetrate deeper into the ambient water, they must rise correspondingly higher at their top above the surface level of adjacent ambient water. This effect creates a hill of pool water that is flat at its highest elevation above regions where the lower interface is itself level, but that slopes downward into the frontal zone to meet the ambient water surface as the frontal interface slopes upward there. This geometry is illustrated in Figure 3.9. Since the actual density fractional difference between pool water and ambient is only slight (around 10^{-2}), the elevation of the free surface above ambient is correspondingly slight compared with the interfacial depth ($\eta/D \sim \Delta\rho/\rho_\infty \ll 1$). Nevertheless this downward free surface slope is sufficient to push near-surface water "downhill" toward the front. When it arrives there it cannot cross the frontal interface because the large stability of the vertical stratification resists mixing of the water masses. Instead, this water must rapidly sink there and then be dragged away by means of internal friction arising from the motion of ambient water below, which has been forced to flow downward beneath the pool. The overall balance of mass and momentum in these processes selects a value for u_∞, the approach speed of ambient water (or, alternatively, the speed of frontal propagation as seen by an observer moving with ambient water). Quite analogous dynamics govern the motion of a tidal bore—a large-amplitude, breaking gravity wave in shallow water; in fact, an estuary front is simply a strong, breaking *internal* gravity wave.

A model of the dynamics is presented in detail in Garvine (1974b). Figure 3.10 shows one streamline pattern computed from that model with flow parameters matched to those of the Connecticut River plume front. The surface convergence is clear in this flow pattern, as is the strong resultant sinking motion. Within the light-water pool a recirculation cell exists, which could be highly effective in promoting lateral circulation across the plume.

Estuary fronts are not isolated physical phenomena. They have larger-scale counterparts, such as the Gulf Stream

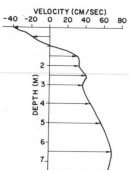

FIGURE 3.8 Horizontal velocity component normal to the front versus depth for the Connecticut River plume 30 m from the front. From Garvine and Monk (1974).

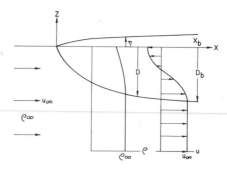

FIGURE 3.9 Configuration of the frontal zone showing typical profiles of density ρ and horizontal velocity u. The interfacial depth is D, and free surface elevation is η.

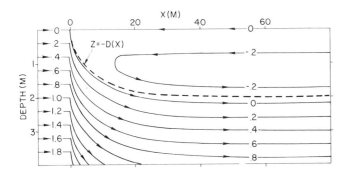

FIGURE 3.10 Computed streamlines from the model of Garvine (1974b). The dashed line indicates the depth of the light-water pool. Arrows show direction of flow.

front and the equatorial front in the Pacific explored by Beebe (1926). Rotation of the earth often plays an important role in the dynamical motion of these larger oceanic fronts.

SUGGESTED RESEARCH AREAS

The body of scientific literature on river plumes and estuary fronts contains papers that largely document the distribution of properties. Observations of the mean motion field are sparse. Therefore, it is difficult to comprehend the dynamics behind observed distributions of properties and hazardous to construct theoretical models of these phenomena. As a prelude to further progress in understanding and modeling these phenomena, we need trustworthy observations that adequately sample the mean motion both in time and space. This task is the more arduous, since time and space scales for estuarine dynamics are comparatively short. Not all observations need be made in the field, however. For many of the phenomena, laboratory models, properly scaled dynamically, should be highly useful.

Together with observations of the mean motion field, we are quite lacking in a necessary understanding of the turbulence characteristics of estuaries and fronts. In particular, our knowledge of turbulent flows in stratified fluids, such as predominate in estuaries, is weak. Appeal to uniform mixing coefficients in theoretical models will not suffice. Turbulent transport of mass, momentum, salt, and heat are characteristics of a flow, not of a fluid. We need to know more about such flows.

REFERENCES

Beebe, W. (1926). *The Arcturus Adventure*, G. P. Putnam's Sons, New York and London, Chap. II.

Garvine, R. W. (1974a). Physical features of the Connecticut River outflow during high discharge, *J. Geophys. Res. 79*, 831.

Garvine, R. W. (1974b). Dynamics of small-scale oceanic fronts, *J. Phys. Oceanog. 4*, 557.

Garvine, R. W. (1975). The distribution of salinity and temperature in the Connecticut River Estuary, *J. Geophys. Res. 30*, 1176.

Garvine, R. W., and J. D. Monk (1974). Frontal structure of a river plume, *J. Geophys. Res. 79*, 2251.

Hansen, D. V., and M. Rattray, Jr. (1966). New dimensions in estuary classification, *Limnol. Oceanog. 11*, 319.

Ingram, R. G. (1976). Characteristics of a tide-induced estuarine front, *J. Geophys. Res. 81*, 1951.

Szekielda, K.-H., S. L. Kupferman, V. Klemas, and D. F. Polis (1972). Element enrichment in organic films and foam associated with aquatic frontal systems, *J. Geophys. Res. 77*, 5278.

Wright, L. D., and J. M. Coleman (1971). Effluent expansion and interfacial mixing in the presence of a salt wedge, Mississippi River delta, *J. Geophys. Res. 76*, 8649.

Fjord and Salt-Wedge Circulation

4

MAURICE RATTRAY, JR.
University of Washington

INTRODUCTION

Three kinds of estuary can be identified by their geomorphology (Cameron and Pritchard, 1963). Coastal-plain estuaries are essentially wide rivers, flowing across coastal plains into the sea. They are usually shallow and may have a dredged channel for navigational purposes. Examples are Chesapeake Bay on the East Coast, the Mississippi River on the Gulf Coast, and the Columbia River on the West Coast. Fjords are deep, glacially scoured inlets, which in many cases have sills at their mouths. Examples are Hood Canal, Silver Bay, and many inlets in British Columbia and southeast Alaska, as well as in Norway and Chile. And finally, there are bar-built estuaries in which bays or lagoons are separated from the open sea by bars through which there may be one or more channels. These occur commonly on the Gulf Coast. In many ways, they behave more like a small sea or a lake than they do like either a coastal plain or the fjord estuary; that is, they will have an important lateral circulation and relatively minor effects due to the varying density of the waters. In this chapter only the behaviors of fjord estuaries and salt wedges in the coastal-plain estuary are considered.

The time-average estuarine circulation forms a complicated dynamic system in which the mutual interaction of the current and the density fields must be considered. A classification scheme based on the variations in distribution of properties in estuaries depends on two parameters: a measure of the circulation and a measure of the stratification.

Different estuarine regimes occur for different values of these parameters, as shown in Figure 4.1. An important research objective, met only for simple coastal-plain estuaries, is to be able to locate a given estuary on this stratification-circulation diagram from a knowledge of its bulk parameters: the river runoff, the depth of the estuary, the tidal current speeds, and the density difference between the source seawater and freshwater. Seven types of estuary are identified by use of the stratification-circulation diagram. Type 3, typical of fjords, is distinguished primarily by the dominance of advection accounting for over 99 percent of the upstream salt transfer. In Type 3b estuaries the lower layer is so deep that in effect the salinity gradient and the circulation do not extend to the bottom, an important qualitative difference from other types of estuary. Fjord estuaries are generally of Type 3b until mixed to the extent

36

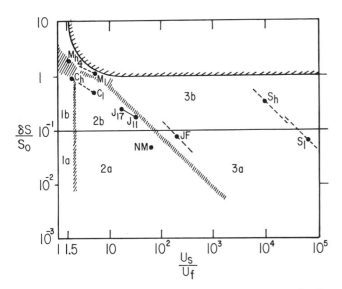

FIGURE 4.1 Proposed classification with some examples. Station code: M, Mississippi River mouth; C, Columbia River Estuary; J, James River Estuary; NM, Narrows of the Mersey Estuary; JF, Strait of Juan de Fuca; S, Silver Bay. Subscripts h and l refer to high and low river discharge; numbers indicate distance (in miles) from the mouth of the James River Estuary. From Hansen and Rattray (1966).

that they assume the Type 3a characteristics with small stratification. In Type 4, salt-wedge estuaries, the stratification is still greater, with an upper layer flowing out over a bottom layer that penetrates upstream in the form of a wedge.

FJORDS

GEOMORPHOLOGY

Fjords are deep embayments formed by glacial scouring usually with moraines forming sills within or at their entrance. Fjords are long, of relatively constant width, and with steep sides. Rivers may discharge into the head of the fjord or at one or more locations along its sides. Figures 4.2 and 4.3 illustrate these features.

STRATIFICATION

Fjords exhibit a great range in their stratification, varying from strong stratification in the surface layers near river outflow to weak stratification in the deep waters and over sills with strong tidal mixing. Seasonal effects are apparent because of annual cycles in runoff and heating. A few examples will illustrate the nature of these variations.

The classic highly stratified situation is shown in Figure 4.2 by a series of vertical salinity profiles along the axis of Knight Inlet, British Columbia, in July 1956, a time of large runoff (Pickard, 1956). There is a marked stratification in the upper 10 to 20 m or so within the inner basin,

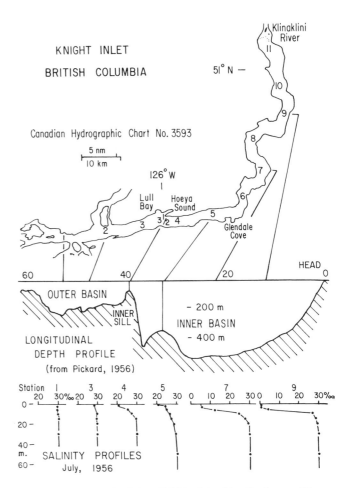

FIGURE 4.2 Knight Inlet, British Columbia: Station positions, longitudinal profile, and salinity-depth profiles. From Pickard and Rodgers, 1959.

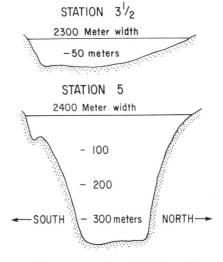

FIGURE 4.3 Transverse section at Stations 3½ and 5, Knight Inlet. From Pickard and Rodgers (1959).

which diminishes toward the mouth because of mixing and upward advection of denser, more saline water from below. The deep waters of the basin are weakly stratified everywhere. A typical seasonal cycle for salinity and temperature is shown in Figures 4.4 and 4.5 from Prince William Sound, Alaska (Muench and Schmidt, 1975). It is clear that the annual cycles of heating and runoff do not coincide and that the temperature profile may show inversions with a stable salinity-controlled density profile. At high latitudes, in the winter season, runoff may cease and vertical thermohaline convection extend essentially to the bottom, as illustrated in Figure 4.6.

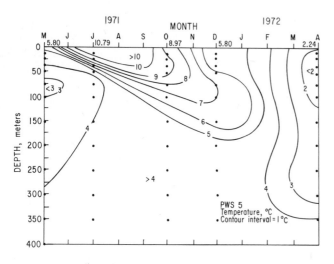

FIGURE 4.4 Time series of temperature at PWS 5 from May 1971 to April 1972. From Muench and Schmidt (1975).

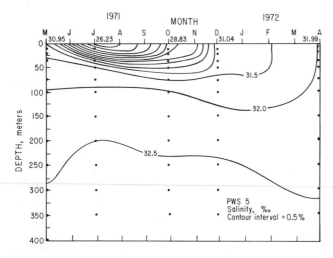

FIGURE 4.5 Time series of salinity at PWS 5 from May 1971 to April 1972. From Muench and Schmidt (1975).

CIRCULATION

The mean circulation under conditions of large runoff and small tidal mixing is restricted to the upper layer and consists of a strong outflow in a very shallow layer and an inflow immediately underneath. With weak stratification or strong tidal mixing over sills, the circulation penetrates deeper and may reach the bottom to become similar to that which occurs in coastal-plain estuaries. These features are shown in Figures 4.7 and 4.8 (Rattray, 1967).

The effects of wind have been difficult to determine because of their transient nature. However, several observations of reasonably steady winds lasting for more than one day have shown that up-inlet winds can retard the surface outflow and may exert sufficient stress to reverse it, forcing the outflow to occur in an intermediate layer just below the surface, as seen in Figure 4.9 (Pickard and Rodgers, 1959). When the runoff is slight and the stratification weak, a three-layer wind-driven circulation can extend to the bottom, as shown in Figure 4.10 (Rattray, 1967).

DYNAMICS OF THE MEAN CIRCULATION

The dynamic balance is between the horizontal pressure gradient and the vertical gradient of turbulent stress, except for the upper layer circulation, where the nonlinear field accelerations are also important. The salt (density) balance is between advection and vertical diffusion. Under these conditions, the governing equations become, in dimensionless form (Winter, 1973),

$$\delta \frac{\partial}{\partial z} \left(u \frac{\partial u}{\partial x} + w \frac{\partial u}{\partial z} \right) = \frac{\partial^2}{\partial z^2} \left(K \frac{\partial u}{\partial z} \right) - Rf \frac{\partial \Sigma}{\partial x},$$

$$(4.1)$$

$$u \frac{\partial \Sigma}{\partial x} + w \frac{\partial \Sigma}{\partial z} = \frac{\partial}{\partial z} \left(K \frac{\partial \Sigma}{\partial z} \right), \qquad (4.2)$$

$$\frac{\partial bu}{\partial x} + \frac{\partial bw}{\partial z} = 0, \qquad (4.3)$$

$$\frac{\rho}{\rho_0} = 1 + \epsilon \frac{S}{S_0}, \qquad (4.4)$$

where the following characteristic values have been chosen:

$\sigma_0 = \delta S/S_0$, the relative salinity excursion;
$u_0 = R_0/b_0 z_0 \sigma_0$, the horizontal velocity scale;
$w_0 = u_0 z_0/x_0$, the vertical velocity scale;
$x_0 = u_0 z_0^2/K_0$, the horizontal length scale;
z_0 is a characteristic vertical length;
K_0 is a characteristic vertical eddy diffusivity;
b_0 is a characteristic width;
R_0 is a characteristic value of runoff;
$\delta = K_z/N_z$, a ratio of vertical eddy diffusivity to vertical eddy viscosity, assumed constant;
$\tau_0 = \rho_0 u_0 K_0/z_0 \delta$, the characteristic wind stress;
$R_f = \delta g \epsilon \sigma_0 z_0/u_0^2$, the overall flux Richardson number.

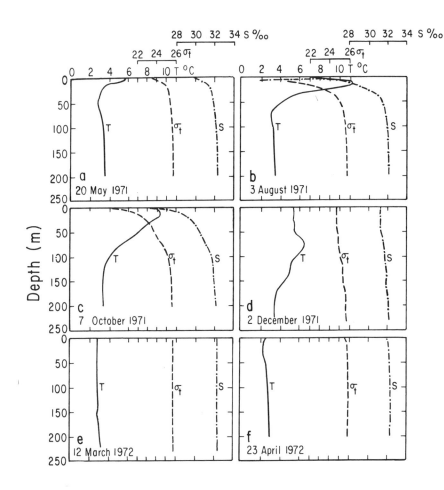

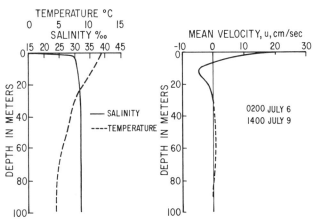

FIGURE 4.6 Vertical profiles of temperature, salinity, and σ_t (density) at Station 130 in Port Valdez for six different times during the period May 1971–April 1972. From Muench and Nebert (1973).

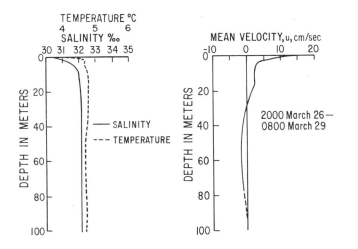

FIGURE 4.7 Temperature, salinity, and velocity profiles, Silver Bay, March 1957. From Rattray (1967).

FIGURE 4.8 Temperature, salinity, and velocity profiles, Silver Bay, July 1956. From Rattray (1967).

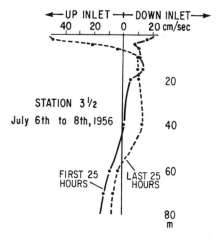

FIGURE 4.9 Net current profiles over first and last 25 hours at Station 3½, Knight Inlet, July 6-8, 1956. From Pickard and Rodgers (1959).

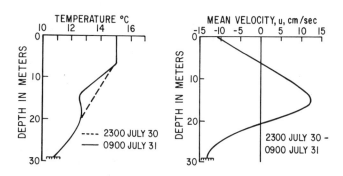

FIGURE 4.10 Temperature and velocity profiles, East Sound, July 1956. From Rattray (1967).

New variables, the streamfunction, ψ, and the salinity defect, Σ, are defined by

$$bu = -(\partial\psi/\partial z), \qquad bw = \partial\psi/\partial x, \qquad (4.5)$$

$$S/S_0 = 1 - \sigma_0\Sigma \qquad (4.6)$$

and are presumed to approach zero at great depth.

The boundary conditions at $z = 0$ are

$$-K\frac{\partial u}{\partial z} = \tau_w, \qquad (4.7)$$

$$\psi(x,0) \doteq 0, \qquad (4.8)$$

the latter applicable when the runoff is small compared to the total circulation. The effect of runoff is obtained from the constraint condition on the total salt flux:

$$\int_0^\infty bu\,\Sigma\,dz = R(x). \qquad (4.9)$$

Rattray (1967) and Winter (1973) have presented similarity solutions to these equations, which retain the non-linearity but require specific relationships for all longitudinal dependencies. The main features of the mean circulation, as shown in Figure 4.11 for Knight Inlet by Winter, are reproduced by this solution. However, the eddy coefficients and their variation with position are chosen for fit and are not determined from the theory. Nevertheless, these solutions can give useful analytic descriptions of the principal features of river- and wind-driven mean flows and density structures for deep fjords.

TIDES AND TIDAL CURRENTS

Because of the great depths of fjords, there is little tidal dissipation, except over the sills, where the currents may be quite turbulent and relatively uniform with depth. This

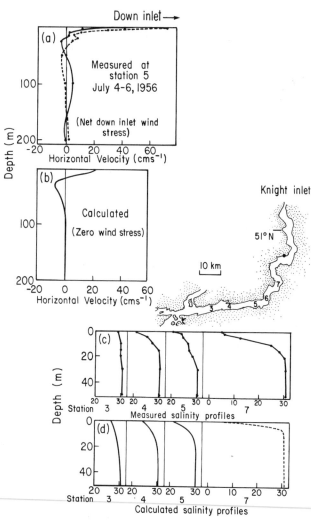

FIGURE 4.11 Comparison of (a) measured and (b) calculated horizontal velocity profiles for Station 5 in Knight Inlet. Comparison of (c) salinity measurements acquired at the four stations indicated in the map of Knight Inlet with (d) calculated depth profiles of salinity. From Winter (1973).

means that the tidal phase will change more rapidly across the sills than along the deeper reaches of a fjord. Correspondingly, the tidal range will increase toward the heads of fjords with the more rapid increase occurring over the sills. Extreme examples of reversing falls or rapids can occur at shallow or constricted entrance sills.

Tidal current behavior in the deep basins can be extremely complicated because of the generation of internal waves at the sills. In some cases, these waves may take the form of an undulating bore propagating away from the sill during the flood current, as observed in some detail by Halpern (1971) in Massachusetts Bay and by Shand (1953) in the Strait of Georgia. Another possibility, not well documented but suggested by the measurements of Pickard and Rodgers (1959), is that internal tides are set up, as first proposed by Zeilon (1913, 1934), by the bathymetric coupling mechanism elaborated by Prinsenberg *et al.* (1974) and by Baines (1973, 1974).

EFFECTS OF VARIABLE WINDS ON FJORD CIRCULATION

Although it has been known since early studies in Gullmarfjord (Sandstrom, 1904; Petterson, 1920) that variable winds have a marked effect on the currents and density structure of fjords, it is only recently that a dynamical description has been presented by Farmer (1972) that explains many features he observed in an intensive observational program in Alberni Inlet. Figure 4.12 clearly shows the

reversal of the current in a shallow layer caused by the onset of a strong up-inlet wind. Figure 4.13 illustrates the corresponding effect on the density structure. The coherence between current and wind is high for periods greater than about 20 hours. The phase angle changes almost logarithmically throughout the range, but at periods longer than 60 hours, the current appears to lead the wind. For periods less than about 40 hours, the phase-lag dependence of the current is consistent with the propagation of a viscous shear wave with constant eddy viscosity from the surface to the 2-m depth.

Typically, Farmer found that strong up-inlet winds caused a sudden thickening of the surface layer at the inlet head. This distortion took at least two or three days to relax. Away from the head, the disturbance was both attenuated and delayed. He explained this behavior through the action of the internal mode of a two-layer system approximating the observed density distribution. The response to a changing wind then consists of damped waves traveling both up and down inlet about the equilibrium configuration. A comparison of the observed and theoretical results is shown in Figure 4.14. The results are remarkably good considering the simplified representation of the system that was used. They do depend on a suitable choice of friction coefficients.

REPLACEMENT OF DEEP WATER BEHIND SILLS

Deep water behind sills generally does not participate in a steady circulation but is usually replaced rapidly during relatively short intervals between longer periods of quiescence. Most commonly, this deep water replacement is related to an annual cycle of density in the source waters above sill depth. Whether replacement occurs every year de-

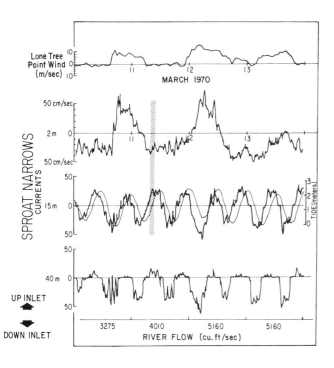

FIGURE 4.12 Wind, current, and mean daily river discharge. Measured tidal elevation at Port Alberni is superimposed on 15-m current. The vertical gray line indicates time of drift pole measurements. From Farmer (1972).

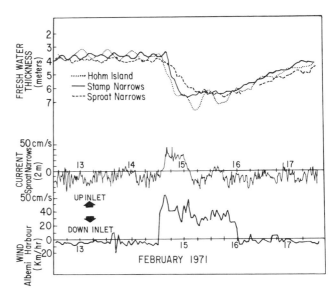

FIGURE 4.13 Freshwater thickness, 2-m current, and wind (5 days). From Farmer (1972).

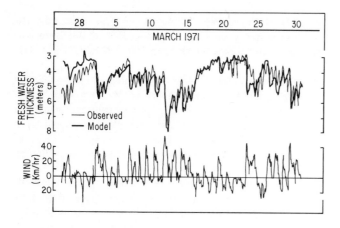

FIGURE 4.14 Comparison of model and data for freshwater thickness at Hohm Island and measured wind speed at Lone Tree Point. Since the model starts with zero initial conditions, it takes a few days to catch up with the data. From Farmer (1972).

pends on the year-to-year variability in maximum source density and on the rate of density reduction in the basin due to diffusive processes. Gade (1973) has studied this replacement mechanism as a stochastic process.

More-complicated situations arise when exchanges across the sill are neither complete nor restricted to short periods of time. Welander (1974) considers this question for a two-layer system. Parameterizing the exchanges between the two layers and considering simple relationships for the exchanges across the sill, he elucidates many features of possible time-dependent salinity fields and the corresponding flushing rate for the deep water behind the sill. His main results show that (1) a single steady state exists that is approached by an exponential decay from an initial state; (2) the total mixing through the interface must decrease with depth of the interface to allow a stable steady state; (3) the static stability increases with increasing freshwater supply, up to a critical value where the two-layer model breaks down; (4) an added oscillatory component in source salinity increases, while in the runoff it decreases, the estuary salinity and static stability; and (5) the statistical steady state is sensitive to certain high-order statistical features of the forcing functions.

When intrusion of deep water with differing properties occurs in small pulses separated by quiescent periods, it is possible to generate a complex structure of both conservative and nonconservative properties in the basin. When this mechanism provides clearly marked parcels of water, their time history can provide information on the currents and turbulent diffusion taking place. Ebbesmeyer et al. (1975) carried out such a study evaluating both the effects of current shear and turbulent diffusion. They found, from adequately sampled large-scale features, the shear to be dominated by intrusion and estimated $K_z \approx 1$ cm^2 sec^{-1} and $K_x \approx 10^5$ cm^2 sec^{-1}.

SALT-WEDGE ESTUARIES

STRATIFICATION AND CIRCULATION

Salt wedge is the term commonly applied to describe the salinity distribution in a highly stratified coastal-plain estuary with nearly homogeneous upper and lower layers separated by a strong halocline. The halocline depth, following the general shape of a wedge, increases with increasing distance up the estuary. Typically, strong outward flow occurs in the upper layer, with a slower recirculation occurring in the bottom layer. These features are shown in the stratification–circulation diagram, along with the conditions favoring existence of a salt wedge: $F_m > 10^{-2}$ and $P > 10^{-1}$, where $F_m = R/[g(\Delta\rho/\rho)H^3]^{1/2}$, $P = R/U_t H$, with R the river discharge per unit width of the estuary, H the depth of the estuary, g the acceleration due to gravity, ρ the density of estuarine water, $\Delta\rho$ the density difference between river water and seawater, and U_t the rms tidal current speed. Thus salt wedges occur with high discharge rates per unit width, shallow depths, and relatively weak tidal currents. An often cited example is the Mississippi River.

With these conditions, the upward salt flux across the halocline is sufficiently small so that the horizontal salinity gradients in the two layers are dynamically unimportant and the pressure field results almost entirely from the slopes of the free surface and the halocline. The salt-wedge regime breaks down if the salt flux through the halocline is increased sufficiently by mixing due to tidal currents, interfacial exchanges from current shear instabilities, and resulting turbulence or, possibly, growth in the surface area of the halocline with large salt-wedge penetration upstream. There is no information directly applicable to the salt-wedge flow regime that gives dependency of the salt flux across the interface on the external parameters of the flow. Possibly the most applicable results have been obtained experimentally by Kato and Phillips (1969) and by Lofquist (1960). These are summarized by Turner (1973), who fits the experiments of Kato and Phillips with the expression

$$\frac{u_e}{u_i^*} = 2.5 \frac{\rho_1 u_i^{*2}}{g(\rho_2 - \rho_1)h_1}, \qquad (4.10)$$

where u_e is the entrainment velocity, u_i^* is the friction velocity at the interface, g is the acceleration of gravity, ρ_1, ρ_2 are the densities in the upper and lower layers, respectively, and h_1 is the depth of the upper layer. The limiting case for small $\rho_1 u_i^{*2}/g(\rho_2 - \rho_1)h_1$ was investigated by Fortescue and Pearson (1967), who explained their results on the basis of molecular diffusion. It can be shown that the entrainment velocity given by Eq. (4.10) will not have a significant effect on the dynamics of a salt wedge.

Field data adequate for a complete description of the stratification and circulation in a salt-wedge estuary are rare. Results from the Ishikari River (Fukushima et al., 1969) are useful in this regard as they not only show the shape and penetration of the wedge but provide vertical profiles of salinity and velocity and the distribution of surface salinity as a function of discharge. Data from a flume calibrated for the Mississippi River (Rhodes, 1950) include the profile of the salt wedge for different river discharges. An example of the change in wedge with change in discharges for the Duwamish River is shown in Figure 4.15 from Dawson and Tilley (1972).

Laboratory experiments (Farmer, 1951; Keulegan, 1957; Shi-Igai and Sawamoto, 1969) have provided the main basis for experimental comparison with theoretical results. Keulegan (1966) summarizes much of this work. Because of the restricted range of Reynolds number that can be achieved in the laboratory, it is uncertain how well the results of these experiments extrapolate to natural systems. However, the data can be compared with theoretical results.

DYNAMICS OF THE MEAN CIRCULATION

Theoretical studies determining the length and profile of salt wedges have been obtained by Sanders *et al.* (1953) and Shi-Igai and Sawamoto (1969). These have been based on assumptions of two-layer flow with no salt exchange across the interface, a constant interfacial stress coefficient, negligible velocities in the lower layer, and negligible bottom stress. While agreement can be obtained between predicted and observed lengths of salt wedges by the choice of a particular value for the interfacial stress coefficients, there is no independent check on the correctness of the assumptions used. The values of the stress coefficient are found to vary with conditions in a particular estuary and from estuary to estuary.

Rattray and Mitsuda (1974) have extended the previous work to include the effects of bottom slope and bottom friction. In addition, they obtain the complete circulation, permitting the interfacial stress to be related directly to the flow conditions. They balance the pressure gradient and turbulent frictional stress in the lower layer and in the upper layer, adding the effect of the field acceleration for the latter. The frictional stress and velocity are taken to be continuous across the interface, with the assumption that the interface and the bottom behave as smooth boundaries for the turbulent flow in each layer. Results compare favorably with laboratory experiments and with data from the Duwamish and Mississippi Rivers. The comparison of the density profiles is given in Figures 4.16 and 4.17. Likewise, the velocity profiles are compared in Figure 4.18.

TIDES AND TIDAL CURRENTS

As a first approximation, the wedge and its bounding halocline will simply be advected by the barotropic tide motions. In addition, nonuniform cross sections will induce tidal deformations in the wedge profile, but as the whole system is strongly frictional, one would not expect significant tidal

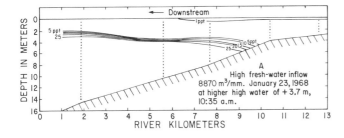

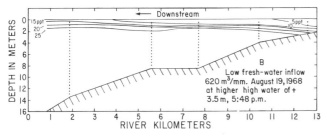

FIGURE 4.15 Schematic longitudinal profile of estuary showing distribution of salinity at high and low rates of freshwater inflow at high tide. Numbered line represents depth distribution of salinity value given in parts per thousand. Dots indicate points of individual measurements, which were made at half-meter depth intervals. From Dawson and Tilley (1972).

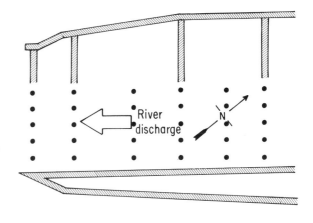

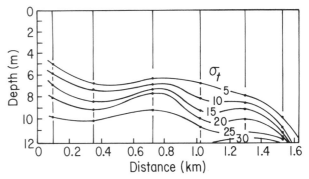

FIGURE 4.16 Density profiles for a salt wedge observed near the mouth of Southwest Pass at a discharge of 6120 m³/sec. Also shown is the plan of the jetty system at the same location. Water samples were taken at 1.5-m (5-ft) depth intervals; station positions are shown as circles in the upper diagram and as vertical lines in the lower diagram. (Data from Sheet 3, Plate II, Corps of Engineers, 1959.) From Rattray and Mitsuda (1974).

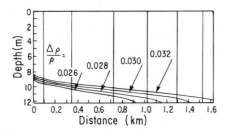

FIGURE 4.17 Computed profiles corresponding to wedge shown in Figure 4.18. Station positions, indicated by vertical lines, have been included for reference. From Rattray and Mitsuda (1974).

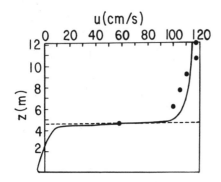

FIGURE 4.18 Comparison of current-meter measurements for a wedge observed in Southwest Pass (from Sheet 1, Plate II, Corps of Engineers, 1959) with the computed velocity profile. Measurements are denoted by circles. The dashed line indicates the depth at which the interface was assumed to be for the purpose of computation. From Rattray and Mitsuda (1974).

internal oscillations to occur. Dawson and Tilley (1972) determined the oscillation of the wedge in the Duwamish River for both large and small tidal ranges. Garvine (1975) has determined the horizontal tidal displacement of the salt wedge in the Connecticut River as a function of P, the ratio of freshwater outflow velocity to tidal velocity.

CURRENT PROBLEMS AND RESEARCH NEEDS

At present, our understanding and prediction capability of fjord and salt-wedge estuarine circulation are limited by two categories of deficiency: (1) inadequate knowledge of the relevant physical processes and (2) the difficulty of calculating and/or observing the actual circulation and transport in an estuary because of the complicated nonlinear behavior, the irregular geometry, and the variable nature of the forcing.

The first category includes the most important question of the momentum and salt-flux mechanisms acting in the highly stratified regions occurring in the upper layers of fjords and at the top of the salt wedge. Further progress on

understanding the related circulations is impossible until these processes can be more adequately described. The fjord and salt-wedge estuaries are ideal for investigation of this question, as they provide a large Reynolds number hydrodynamic laboratory without the logistic and scale difficulties inherent for comparable investigations in the open sea.

Several experiments have been undertaken that directly measure fluxes by correlation methods. However, there has not been an intercomparison experiment to demonstrate the characteristics of each type of current meter compared with the others. It is timely to have such an intercomparison experiment as an important step in the further investigation of the turbulent processes in estuaries.

Also, the mixing mechanisms in the deep basins have not been determined. What is the turbulence level? Is the energy source internal tides, boundary mixing, wind-driven transients, or surface circulation? These questions require answering in order to understand and predict the time history of the basin water properties.

Similarly, the water properties in deep basins depend critically on the nature of the exchange processes across entrance sills. The circulation and mixing over these sills are understood only in a crude sense.

The second category requires development of better methods for integrating the governing equations in conjunction with field programs to test the calculated results and to assess what is important for the mathematical description of the system. Because these questions are complex, the field programs required to provide the necessary answers must become more elaborate and extensive than they have been in the past. It will be necessary to carry out experiments with a space and time coverage seldom achieved to date. We do not know such things as the suspended sediment distribution and its effect on the density flow in salt-wedge estuaries, the coupling between variable wind-driven currents and the surface circulation in fjords, the effects of lateral variability or lateral boundaries on the fjord circulation, and the effect of sloping bottoms and channeling on the salt-wedge circulation.

ACKNOWLEDGMENTS

The author acknowledges the support of the National Science Foundation through Grant DES 74-22711 AO1.

REFERENCES

Baines, P. G. (1973). The generation of internal tides by flat-bump topography, *Deep Sea Res. 20,* 179.

Baines, P. G. (1974). The generation of internal tides over steep continental slopes, *Phil. Trans. R. Soc. London 277,* 27.

Cameron, W. M., and D. W. Pritchard (1963). Estuaries, in *The Sea,* Vol. II, John Wiley and Sons, New York, pp. 306–324.

Dawson, W. A., and L. J. Tilley (1972). Measurement of salt wedge excursion distance in the Duwamish River Estuary, Seattle, Washington, by means of the dissolved-oxygen gradient, *Geologi-*

cal Survey Water-Supply Paper 1873-D, U.S. Department of the Interior, Washington, D.C.

Ebbesmeyer, C. C., C. A. Barnes, and C. W. Langley (1975). Application of an advective-diffusive equation to a water parcel observed in a fjord, *Estuarine Coastal Marine Sci 3*, 249.

Farmer, D. M. (1972). The influence of wind on the surface waters of Alberni Inlet, Ph.D. Thesis, Institute of Oceanography, U. of British Columbia.

Farmer, H. G. (1951). An experimental study of salt wedges, Woods Hole Oceanographic Institution Tech. Rep. No. 51-99, Woods Hole, Mass.

Fortescue, G. E., and J. R. A. Pearson (1967). On gas absorption into a turbulent liquid, *Chem. Eng. Sci. 22*, 1163.

Fukushima, H., I. Yukuwa, and S. Takehashi (1969). Salinity diffusion at the interface of stratified flow in estuary, *Proc. of 13th Cong. Internat. Assoc. Hydraulic Res. 3*, 191.

Gade, H. G. (1973). Deep water exchanges in a sill fjord; a stochastic process, *J. Phys. Oceanog. 3*, 213.

Garvine, R. W. (1975). The distribution of salinity and temperature in the Connecticut River Estuary, *J. Geophys. Res. 80*, 1176.

Halpern, D. (1971). Observation on short-period internal waves in Massachusetts Bay, *J. Marine Res. 29*, 116.

Hansen, D. V., and M. Rattray, Jr. (1966). New dimensions in estuary classification, *Limnol. Oceanog. 11*, 319.

Kato, H., and O. M. Phillips (1969). On the penetration of a turbulent layer into stratified fluid, *J. Fluid Mech. 37*, 643.

Keulegan, G. H. (1957). Eleventh progress report on model laws for density currents from characteristics of arrested saline wedges, reprinted from *National Bureau of Standards Report No. 5482*, NBS, Washington, D.C.

Keulegan, G. H. (1966). The mechanism of an arrested saline wedge, in *Estuaries and Coastline Hydrodynamics*, McGraw-Hill Book Co., New York, pp. 546-574.

Lofquist, K. (1960). Flow and stress near an interface between stratified liquids, *Phys. Fluids 3*, 158.

Muench, R. D., and D. L. Nebert (1973). Physical oceanography, in *Environmental Studies of Port Valdez*, Inst. of Marine Science, U. of Alaska, Occasional Publ. No. 3, Fairbanks, Alaska, pp. 103-149.

Muench, R. D., and G. M. Schmidt (1975). Variations in the hydrographic structure of Prince William Sound, Inst. of Marine Science, U. of Alaska Rep. R75-1.

Petterson, H. (1920). Internal movements in coastal waters and meteorological phenomena, *Geograf. Ann. Stockholm I*, 32.

Pickard, G. L. (1956). Surface and bottom currents in the Strait of Georgia, *J. Fisheries Res. Board of Canada 13*, 581.

Pickard, G. L., and K. Rodgers (1959). Current measurements in the Knight Inlet, British Columbia, *J. Fisheries Res. Board of Canada 16*, 635.

Prinsenberg, S., W. Wilmot, and M. Rattray, Jr. (1974). Generation and dissipation of coastal internal tides, *Deep Sea Res. 21*, 263.

Rattray, M., Jr. (1967). Some aspects of the dynamics of circulation in fjords, in *Estuaries*, American Association for the Advancement of Science, Washington, D.C., pp. 52-62.

Rattray, M., Jr., and E. Mitsuda (1974). Theoretical analysis of conditions in a salt wedge, *Estuarine Coastal Marine Sci. 2*, 375.

Rhodes, R. F. (1950). Effect of salinity on current velocities, in *Evaluation of Present State of Knowledge of Factors Affecting Tidal Hydraulics and Related Phenomena*, U.S. Army Corps of Engineers, Vicksburg, Miss., pp. 41-100.

Sanders, J. L., L. C. Maximon, and G. W. Morgan (1953). On the stationary "salt wedge"—a two layer free surface flow, Brown U. Tech. Rep. No. 1, Providence, R.I.

Sandstrom, A. (1904). Publications de Circonstance, No. 18, Copenhagen, quoted in Petterson (1920).

Shand, J. A. (1953). Internal waves in Georgia Strait, *Trans. Am. Geophys. Union 34*, 849.

Shi-Igai, H., and M. Sawamoto (1969). Experimental and theoretical modeling of saline wedges, *Proc. 13th Congress Internat. Assoc. Hydraulic Res. 3*, 29.

Turner, J. S. (1973). *Buoyancy Effects in Fluids*, Cambridge U. Press, Cambridge.

Welander, P. (1974). Two-layer exchange in an estuary basin, with special reference at the Baltic Sea, *J. Phys. Oceanog. 4*, 542.

Winter, D. F. (1973). A similarity solution for steady-state gravitational circulation in fjords, *Estuarine Coastal Marine Sci. 1*, 387.

Zeilon, N. (1913). On the seiches of Gullmar Fjord, *Sven. Hydrogr. Biol. Komm. Skr. 5*, 1.

Zeilon, N. (1934). Experiments on boundary tides, *Goteborgs Kungl. Vetensk. Och Vitter Hetssamhaelles Handiger, 5, B*, No. 3, 1.

Turbulent Processes in Estuaries

5

KENNETH F. BOWDEN
University of Liverpool

INTRODUCTION

The flow in estuaries is nearly always turbulent, and this characteristic affects the mechanics of the flow and its dispersive effects. In one sense, therefore, all physical processes in estuaries could be described as turbulent. In this chapter, a more restricted view will be taken and those aspects considered in which the turbulent nature of the process is apparent and should be explicitly taken into account. These include the turbulent velocity fluctuations themselves, how they may be measured, their structure and spectral distribution, and the Reynolds stresses and eddy transports that they produce. It is of interest to compare these turbulent features with those observed in the laboratory and in other geophysical flows, with a view to applying results from other fields to turbulence in estuaries. Another aspect is how the processes may be related to properties of the mean flow, which are more easily measurable. This involves a parameterization of the turbulent processes in terms of larger scale features, an essential step in modeling estuary circulation and mixing. Progress in parameterization, usually involving eddy coefficients of viscosity and dif-

fusion or dispersion, has been largely empirical in the past, and empirical methods will no doubt continue to be needed. More rapid progress may be made, however, if the physical nature of the turbulent processes is better understood.

Turbulent processes in estuaries may conveniently be divided into three types: (1) those originating at or near the estuary bed and propagating upward; (2) those in the interior of the fluid, i.e., at middepths, where density gradient effects are likely to be more important; and (3) those originating at or near the free surface, e.g., the effect of wind stress in generating waves and a surface drift.

Processes originating at the surface are by no means negligible in estuaries but are generally less important than they are in open coastal waters. For this reason and because surface processes, such as wave generation and breaking and air–sea interaction, are specialized subjects often studied in other contexts, this chapter will concentrate on the first two types: processes originating in the bottom boundary layer and those occurring at middepths. The discussion will be concerned primarily with those coastal-plain estuaries that are partially mixed and in which the tidal flow plays an important part. Salt-wedge estuaries and fjords will not be considered, except in passing.

DIRECT MEASUREMENT OF TURBULENT VELOCITY FLUCTUATIONS

BASIC IDEAS

It is a well-known feature of turbulent flow that the velocity measured at a given point fluctuates in a largely random manner and that the fluctuations are three dimensional even if the averaged flow is unidirectional. If rectangular axes are taken with x along the estuary, y across it, and z vertically upward, and the corresponding components of velocity, measured at a given instant, are u, v, w, then u may be written

$$u = \bar{u} + u', \qquad (5.1)$$

where $\bar{u}$ is the mean value of u over some chosen interval of time and u' is the instantaneous deviation from the mean. Similarly $\bar{v}$, $\bar{w}$ may be defined as the mean velocity components and v', w' as the turbulent velocity components in the y and z directions. By definition,

$$\overline{u'} = \overline{v'} = \overline{w'} = 0. \qquad (5.2)$$

The mean-square quantities $\overline{u'^2}$, $\overline{v'^2}$, $\overline{w'^2}$ may be termed the components of turbulent intensity. The total turbulent intensity is given by

$$\overline{q^2} = \overline{u'^2} + \overline{v'^2} + \overline{w'^2}, \qquad (5.3)$$

and $\frac{1}{2}\rho\overline{q^2}$ is the additional kinetic energy of the flow due to the turbulence.

It was shown by Reynolds in his fundamental work on turbulence that the presence of the velocity fluctuations gives rise to additional pressure and shear stresses in the fluid. Across a horizontal element of area, for example, perpendicular to the z direction, the Reynolds shear stress in the x direction by which the water above the area acts on the water below may be denoted by τ and is given by

$$\tau = -\rho\overline{u'w'}, \qquad (5.4)$$

where ρ is the density of the water. A finite value of $\overline{u'w'}$ implies a correlation between the u' and w' fluctuations.

MEASUREMENTS

One of the first attempts to measure turbulent fluctuations of velocity in an estuary was that of Francis *et al.* (1953) in the Kennebec Estuary, Maine. They used a propeller-type current meter for measuring velocity and a tilting vane for measuring the inclination of the flow and hence the vertical component of velocity. The mean product $\overline{u'w'}$ was computed, but the stresses found were surprisingly high and variable. The records were read at 3-sec intervals over a length of 2 min, so that the variability is not surprising in view of what we now know about sampling frequency and duration of records. They also recorded temperature fluctu-

ations with a thermistor, estimated coefficients of eddy viscosity and diffusion, and attempted to relate them to a Richardson number, as the estuary was slightly stratified. This investigation was remarkable not for its results but for the foresight shown in its planning and the attempt to measure all the relevant parameters.

At about the same time, measurements were made in the Mersey Estuary (Bowden and Fairbairn, 1952) of the longitudinal velocity fluctuations u' only, using the Dodson current meter, which has a propeller that does not rotate freely but is turned through a limited angle against a spring system, with the angle of deflection directly proportional to speed. Two current meters, with either horizontal or vertical spacing, were mounted in a stand laid on the bottom, so that some information on the scale of the fluctuations was obtained. The advantage of the spring-limited propeller was a reduction in the time constant so that fluctuations with periods down to about 1 sec were recorded.

Other investigations have been made with propeller-type current meters, but it is difficult to reduce their time constant much below 1 sec so that fluctuations of frequencies above about 1 Hz are not recorded. In order to get a better time response, an electromagnetic flowmeter system with no moving parts and capable of measuring all three components of turbulent velocity was developed and used, first in coastal waters and later in the Mersey Estuary (Bowden and Fairbairn, 1956; Bowden and Howe, 1963). Electromagnetic current meters have been much improved in recent years and are capable of responding to frequencies up to 10 Hz and resolving distance scales down to about 5 cm. Higher-frequency response is possible using the acoustic Doppler shift current meter, which has been developed at the Chesapeake Bay Institute (Wiseman *et al.*, 1972). This instrument has a frequency response up to at least 20 Hz and examines a volume of water less than 2 cm^3 in volume. Hot-wire instruments, widely used in laboratory work on turbulence, have been adapted for use in the sea, particularly by Grant *et al.* (1962), who have used a hot-film instrument to measure turbulence in the *Discovery* passage. These instruments have a much higher-frequency response, up to several hundred hertz, and can be used in the inertial subrange region of the turbulence spectrum.

The general character of the spectra of the turbulent velocity fluctuations in an estuary is shown in Figure 5.1, taken from a paper by Seitz (1973) and based on measurements in the Patuxent Estuary, flowing into Chesapeake Bay. The scale of the vertical fluctuations w' is limited by the presence of the bottom and surface boundaries so that the spectral intensity falls off at the lower frequencies. On the other hand, the longitudinal fluctuations u' are not limited in this way, and the spectral intensity remains large at the lower frequencies. The spectral behavior of the transverse fluctuations v' resembles that of u' but at a lower level of intensity. The rms value of the u' fluctuations near the bottom is of the order of 10 percent of the mean current, and that of the v' and w' fluctuations somewhat lower.

The spectral distribution of $\overline{u'w'}$, which is also the spectrum of the horizontal shearing stress, reaches a peak at in-

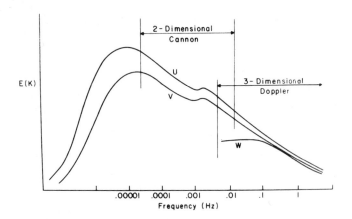

FIGURE 5.1 Energy spectra of *u, v,* and *w* components in a tidal estuary. From Seitz (1973).

termediate frequencies or wavenumbers, which may be regarded as typical of the scale of the turbulence generating process. At lower wavenumbers there is little energy in the vertical fluctuations, w', whereas at the higher wavenumbers the correlation between u' and w' becomes small as the inertial subrange is approached and the turbulence becomes increasingly isotropic. The overall correlation coefficient between the u' and w' fluctuations is negative, and its magnitude is 0.3 to 0.4. To record the main energy-containing eddies and measure the shearing stress by the eddy correlation technique, it is probably adequate to use flowmeters of the electromagnetic or acoustic type with frequency response up to 10 or 20 Hz. To record turbulence in the dissipative subrange, however, and use the spectral curve to derive the rate of energy dissipation, instruments of the hot-wire or hot-film type would be needed.

The above properties of turbulent flow within a few meters of the bed in estuaries or coastal waters are comparable with those of turbulence measured in experimental channels or in the atmospheric boundary layer. This applies to the intensities of the turbulent fluctuations relative to the mean current, the spectral distributions of the fluctuations, and the shear stress.

INTERMITTENCY

An important feature that is well known in other types of turbulent flow is its intermittent character. This feature, which has been much studied in wind-tunnel and water-channel experiments, has also been recorded in flow near the seabed. A review of intermittency in large-scale turbulent flows was given by Mollo-Christenson (1973). It is characteristic of the flow near a boundary that the generation of turbulence and the Reynolds stresses associated with it occur in bursts or events that occupy a relatively small proportion of the total time but in which the stresses reach magnitudes many times greater than their mean values, averaged over a longer period.

Such events were reported by Gordon (1974) in measurements in the Choptank River, Chesapeake Bay, using a pivoted vane current meter. An example is shown in Figure 5.2, which indicates the pronounced intermittency in the contributions of $u'w'$ to the Reynolds stress. From 100 such records, each 8 min long and sampled at 2-sec intervals, 60 percent of the Reynolds stress was produced by events occupying only 10 percent of the time. Similar results were obtained by Heathershaw (1974) not in an estuary but in the open waters of the Irish Sea. From his data, 57 percent of the stress occurred in 7 percent of the time during events in which the stress exceeded twice the rms value. Individual events had a duration of 5 to 10 sec, and the recurrence interval between events ranged from 20 to 100 sec. Heathershaw used electromagnetic current meters with a frequency response up to 10 Hz, but it is interesting to note that he estimated that nearly all of the Reynolds stress was contributed by frequencies lower than 0.6 Hz.

The occurrence of intermittent events clearly has implications for the measurement of shear stress in estuaries. There is the sampling problem of how long an averaging period should be taken in order to obtain a meaningful value for the stress at the bed, whether by the eddy-correlation technique or the mean-profile method. A further point is how one can define a threshold of stress for bringing material into suspension if, during events lasting 5–10 sec, stresses up to 50 times the mean value may occur.

OSCILLATORY CHARACTER OF THE FLOW

Most laboratory investigations of turbulence have taken place in flows that are maintained in as steady a state as possible. Tidal flows in estuaries are essentially oscillatory in character, and this is a feature that should be taken into

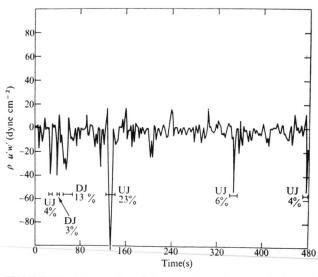

FIGURE 5.2 Time series of the correlation between horizontal (u') and vertical (w') current velocity fluctuations at 2.25 m above bottom showing intermittent, large contributions to the Reynolds stress. From Gordon (1974).

account. Although the time scale for the generation and decay of turbulence is small compared with the tidal period, there are several reasons why the distribution of turbulence and shear stress during a tidal cycle should not be regarded as a series of steady states, each referred to the mean flow at the time: The tidal wave in an estuary is distorted so that the flood currents are usually stronger than on the ebb, and the phase of the current varies with depth. The density current flow is superimposed, although not linearly, on the vertical profile due to the tidal current. Thirdly, the changing pressure gradients and accelerations cause the stress distribution, and probably also the turbulence, to differ in the accelerating and decelerating phases of the flow.

BOTTOM BOUNDARY LAYER

DETERMINATION OF SHEAR STRESS AT THE BED

Turbulence in the bottom boundary layer is directly responsible for the shearing stress at the bed, which is an important parameter in the dynamics of the flow throughout the depth. It also determines the ability of the flow to bring bed material into suspension and transport it along the estuary. Turbulence in this layer also plays a part in the exchange of chemical substances between the water and the sediment.

As the shear stress at the bed is a fundamental parameter in the dynamics of the flow and in considering the effect of the flow on the bed material, there is much interest in methods of determining it. The most direct, in principle, is that already considered: the simultaneous measurement of the u' and w' fluctuations and the formation of their cross-product. In practice, this method would be difficult to use as a routine, since the vertical fluctuations are less easy to measure than the horizontal and in forming the product the two components must be strictly simultaneous. There is some evidence, however, from both turbulence measurements in the sea and in other flows, that the Reynolds stress may be closely related to the overall turbulent intensity $\overline{q^2} = \overline{u'^2} + \overline{v'^2} + \overline{w'^2}$, which is a somewhat easier quantity to measure than the correlation. It is also possible that the relation between the three components may be sufficiently constant for a measurement of the horizontal intensity $\overline{u'^2}$ alone, which can be done with a one-component instrument, to be adequate for estimating the stress.

An alternative method of determining the bottom stress, which has been used widely, is from the profile of mean velocity in the first few meters above the bed. This makes use of one of the best known results in fluid mechanics, the logarithmic velocity law,

$$u = \frac{u^*}{k_0} \log_e \frac{z + z_0}{z_0},$$ (5.5)

where u^* is the friction velocity, defined by $(\tau/\rho)^{1/2}$; k_0 is von Kármán's constant, numerically equal to 0.41; and z_0 is

a roughness parameter, related to the height of the roughness elements on the bed. The validity of the law has been well established in cases where the flow is steady, the fluid is homogeneous, the roughness of the bed is uniform, and the velocity is averaged over a sufficient length of time for intermittent effects to be taken into account. Since none of these conditions is likely to be strictly true in estuary flow, it is not surprising that different workers have come to different conclusions about the adequacy of the logarithmic law. The usual method is for the mean current to be measured at three or four levels, a logarithmic law fitted, and, if the fit is reasonably good, the values of u^*, and hence bottom stress, and also z_0 may be determined.

No systematic study appears to have been made comparing the stress deduced from the logarithmic profile with that measured by the eddy-correlation technique at the same time. It is highly desirable that such experiments be made under a variety of conditions so that the reliability of the profile method may be better assessed.

A further method that is often used in practice is to relate the bottom stress to the mean velocity measured at a single height above the bottom. The equation usually used is the quadratic one:

$$\tau = k\rho\,\overline{u}\,|\overline{u}|,$$ (5.6)

where $\overline{u}$ is measured at a standard height, often 1 m above the bed. k is a friction coefficient that depends on the roughness of the bed and the standard height at which $\overline{u}$ is measured. k has a value of the order of 2.5×10^{-3}, but its exact value depends on the roughness of the bottom and has to be estimated independently for this method to be used.

EFFECT OF MATERIAL IN SUSPENSION

The effect of turbulence in bringing material into suspension and transporting it along the estuary has been mentioned. Another aspect is the reaction of suspended material on the dynamics of the flow. There is evidence from flume experiments that the presence of particles such as sand in suspension causes a damping of the turbulent motion and a modification of the velocity profile corresponding to a given bottom stress. To a first approximation, it appears that the profile, in the steady state, is still logarithmic but with a reduced value of the von Kármán constant, k_0. From observations in tidal currents bearing sand in suspension in some areas of the English Channel and the North Sea, McCave (1973) has proposed a method for modifying the von Kármán constant in these conditions, but he has pointed out that existing theory does not account adequately for all features of the velocity distributions.

If cohesive sediment such as a suspension of clay particles is present, laboratory experiments have indicated that a further modification of the turbulent flow and stress in the boundary layer occurs. This effect has been observed in turbulent tidal flow in a channel at the Island of Nordstrand, North Sea.

TURBULENT FLOW AT MIDDEPTHS

CHARACTER OF THE FLOW

In homogeneous water, one would expect turbulence at all depths, like the mean velocity profile, to be dominated by the bottom stress and the surface gradient. In steady flow in an open channel, it is well known that the internal shear stress increases linearly from the surface to the bottom and the velocity profile has a universal form. In tidal flow, the stress and velocity profiles near maximum flood and ebb approximate those of steady flow, but at other times the acceleration effects result in considerable deviation from these profiles.

In an estuary, the river flow is superimposed on the tidal current, but to a first approximation its effect on the stress and velocity profiles may be regarded as a simple displacement, so that the tidally averaged velocity has a downstream component at all depths, with a corresponding stress component. The horizontal density gradient, which is an essential feature of an estuary, has a more fundamental effect in causing a depth-varying flow, seaward in an upper layer and landward below, to be superimposed on the tidal current. If the vertical density gradient remains negligible, one might expect the vertical distribution of stress and turbulent energy to be only slightly affected. In other words, the density current could be treated as a small perturbation of the tidal flow, and this is a method that has been used for a well-mixed estuary.

However, an estuary flow that varies with depth tends to produce density differences along a vertical by advection. The turbulent fluctuations themselves tend to reduce these differences, but it is doubtful whether, even in a "well-mixed" estuary, the vertical density differences can ever be neglected completely. This point is seen by considering the changes during a tidal period. When the current is strong in flood and ebb, the turbulent mixing may be adequate to maintain almost complete vertical homogeneity; but near slack water the density current flow has a chance to assert itself and produce a layering of lower-salinity water above water of higher salinity. Averaged over a tidal period, the dispersive effect may be considerable, although the average density gradient is small.

The initial effect of a stable density gradient is to reduce the amplitude of the vertical turbulent movements while having little effect on the horizontal movements. This reduces the overall intensity of turbulence and reacts on the shear stress and the vertical eddy diffusion. In time, a new state of dynamic equilibrium will be established, with a reduced intensity of turbulence, especially in the vertical, and probably increased gradients of density and mean velocity. The increased gradients may tend to be concentrated near middepth, leaving relatively well-mixed layers of lower salinity near the surface and higher salinity near the bottom.

The flux of salt, or of another indicator substance, directly associated with the horizontal component of turbulence, is one term contributing to longitudinal dispersion in

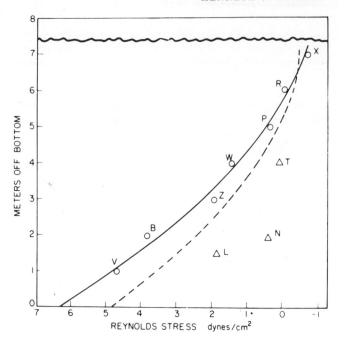

FIGURE 5.3 Measured Reynolds stresses at various depths in the Choptank River Estuary. The letters designate individual data collection runs. The solid line indicates the fit to the points plotted as circles; the dashed line indicates the fit to all points. From Gordon and Dohne (1973).

the estuary. Other terms arise from the vertical variation in current and salinity with both tidally varying and steady components and from similar variations in the transverse direction. In a partially mixed estuary, the longitudinal turbulent flux may account for only a few percent of the total dispersion, as discussed in detail by Fischer (1972). The distribution of salinity in the vertical, however, plays an important part in all these processes, and this distribution is largely determined by the vertical turbulent flux of salt.

Relatively few direct measurements have been made of turbulent fluctuations or Reynolds stresses at middepths. An example from Gordon and Dohne's work (1973) in the Choptank Estuary is shown in Figure 5.3, in which the stress, as determined directly from the $\overline{u'w'}$ correlation, is plotted against depth. All the measurements were made while the speed of the tidal current was near its maximum, and most of the points lie close to a curve that does not deviate greatly from the straight line that would be expected in a steady current. The water was well mixed in this case.

Using the Doppler three-component current meter from a tower in the Patuxent Estuary, Seitz (1973) obtained records of the three components of turbulent velocity, u', v', and w', in conditions for which there was an appreciable density gradient at middepth. The results showed a decrease in the intensity of the vertical fluctuations within a certain spectral band. This band appears to correspond to the buoyancy subrange in which buoyancy forces are extracting energy from the field of turbulence faster than it can be transferred across the energy spectrum by inertial forces.

COEFFICIENTS OF EDDY VISCOSITY AND DIFFUSION

At this stage, it is convenient to introduce some further definitions that are needed in considering these effects quantitatively. Let s denote the salinity or the concentration of some other substance in the water that can be used as a tracer. In the presence of turbulence and of spatial gradients of s, its value at a point will fluctuate and may be divided into a mean and a fluctuating part, as in the case of the velocity components, i.e.,

$$s = \bar{s} + s'. \qquad (5.7)$$

Then, just as the shear stress τ, which is equivalent to the rate of vertical transport of momentum, is given by Eq. (5.4), so the vertical transport of salt due to the turbulence is given by

$$F_z = \rho \, \overline{w's'} \qquad (5.8)$$

per unit area per unit time.

Similarly, the longitudinal transport across a vertical area perpendicular to the flow is given by

$$F_x = \rho \, \overline{u's'}. \qquad (5.9)$$

In order to take the turbulent shear stress into account in the dynamics of the flow, it is necessary to express it in terms of the mean velocity or its derivatives. Prandtl introduced the mixing length l, such that the stress τ is given by

$$\tau = -\rho \, \overline{u'w'} = -\rho l^2 \left| \frac{\partial \bar{u}}{\partial z} \right| \frac{\partial \bar{u}}{\partial z}, \qquad (5.10)$$

and von Kármán considered that l, under certain conditions, is given by

$$l = k_0 \left| \frac{\partial \bar{u}}{\partial z} \middle/ \frac{\partial^2 \bar{u}}{\partial z^2} \right|, \qquad (5.11)$$

where k_0 is von Kármán's constant.

Alternatively, one may define a coefficient of eddy viscosity N_z, analogous to the kinematic molecular viscosity ν, by

$$\tau = -\rho \overline{u'w'} = -\rho N_z \frac{\partial \bar{u}}{\partial z}. \qquad (5.12)$$

Combining Eqs. (5.10) and (5.12), we get

$$N_z = l^2 \left| \frac{\partial \bar{u}}{\partial z} \right|. \qquad (5.13)$$

In this formulation the shear stress at a point is assumed to be determined by the velocity derivatives. This concept

has been questioned by some investigators, e.g., Phillips (1967) and Bradshaw *et al.* (1967), who have deduced that the Reynolds stress at a point is determined by the turbulent field of the whole flow. Although this point of view is probably more accurate, it seems likely that for practical purposes, such as the formulation of mathematical models, the eddy viscosity representation of stress will continue to be used in many cases.

A coefficient of vertical eddy diffusion K_z may be defined in a similar way; thus for the transport of salt in the vertical direction,

$$F_z = \rho \, \overline{w's'} = -\rho K_z \frac{\partial \bar{s}}{\partial z}. \qquad (5.14)$$

In homogeneous flow, the coefficients N_z and K_z may be expected to be approximately equal and to reach their maximum values at middepth. The presence of a stable density gradient tends to reduce them both, but it reduces K_z to a greater extent than N_z so that the ratio K_z/N_z is also decreased. The effect of the stratification on N_z and K_z may be expressed in terms of a suitable parameter, such as the "flux Richardson number," R_f, which is defined by

$$R_f = g(\overline{\rho'w'})/\rho \, (\overline{u'w'}) \frac{\partial \bar{u}}{\partial z}. \qquad (5.15)$$

It is the ratio of the rate of doing work by turbulent mixing against the buoyancy forces to the rate of generation of turbulent energy by the shear stress. Unless there is a significant transfer of turbulence by diffusion into the small region concerned it is necessary that

$$R_f < 1. \qquad (5.16)$$

Equation (5.15), however, contains the turbulent transport terms themselves, and in order to obtain a parameter based on the mean properties of the flow, Eqs. (5.8) and (5.12) are used in Eq. (5.15), giving

$$R_f = \frac{K_z}{N_z} R_i, \qquad (5.17)$$

where R_i is the normal form of the Richardson number,

$$R_i = \frac{g}{\rho} \frac{\partial \bar{\rho}}{\partial z} \middle/ \left(\frac{\partial \bar{u}}{\partial z} \right)^2. \qquad (5.18)$$

The above condition gives

$$R_i < N_z/K_z, \qquad (5.19)$$

Theoretical studies on the instability of small disturbances on laminar, stratified shearing flow have led to the conclusion that the flow becomes turbulent if $R_i < \frac{1}{4}$. It does not necessarily follow that in an initially turbulent flow the turbulence will be completely suppressed when R_i exceeds $\frac{1}{4}$, and there are a number of observations in estuaries, in the open sea, and in other flows indicating that turbulent processes persist at quite high values of R_i.

There are several difficulties in the use of R_i as given by Eq. (5.18), termed the "local Richardson number," as a stability parameter in estuarine flows. Firstly, long-period turbulent components in the velocity u lead to mean values fluctuating considerably from one averaging interval to another. Secondly, the vertical spacing of current meters is often too wide to allow an accurate measurement of gradient $\partial \bar{u} / \partial z$ to be made. A more basic difficulty is that the concept on which R_i is based is that turbulent energy is being generated at the same time and in the same place as it is being used for vertical mixing. If $\partial \bar{u} / \partial z = 0$, it is implied that no turbulent energy is being generated, and R_i becomes infinite. Zero values of $\partial \bar{u} / \partial z$ are bound to occur at some depth at certain stages of the tidal cycle, but it seems unlikely that turbulence will completely disappear at the point concerned. It is more likely that some turbulent energy will be advected or diffused to the point from neighboring regions where $\partial \bar{u} / \partial z \neq 0$. The local Richardson number is, therefore, a parameter to be used with some caution. In view of the difficulties involved in specifying a local R_i, it is sometimes preferable to work with an overall value of R_i, based on larger-scale features of the flow. In a tidal current, for example, an overall R_i may be defined by

$$R_i = gH\Delta\rho / U^2, \qquad (5.20)$$

where U is the depth-mean velocity and $\Delta\rho$ is the total density increase from surface to bottom. There is some theoretical justification for using an overall type of R_i as a turbulent mixing parameter in that the intensity of turbulence and its ability to produce vertical mixing, like the shear stress, is likely to be a property of the flow as a whole, rather than of local gradients of mean velocity and density.

DETERMINATIONS OF STRESSES AND
MIXING COEFFICIENTS

Although few direct measurements of turbulent stresses or turbulent transports of material have been made in estuaries, a number of investigations have been made in which measurements of mean currents and salinity distributions have been used, in conjunction with the equations of motion and of conservation of salt, to derive values of the turbulent transports and the corresponding eddy coefficients. The pioneer work in this field was done by Pritchard (1954, 1956) in the James River, and similar treatments have been given for several other estuaries. It has been confirmed that in relatively well-mixed flow the coefficients N_z and K_z reach their maximum values near middepth and that their magnitudes are reduced by increased density stratification. As the stratification becomes greater at middepths, it causes a greater reduction there in N_z and K_z, so that their vertical distributions become bimodal, with maxima

some distance below the surface and above the bottom but a minimum at middepth.

Various empirical equations have been derived for the dependence of the coefficients N_z and K_z on the Richardson number R_i. A number of these have been discussed by Bowden and Hamilton (1975) and Blumberg (1975) in connection with selecting suitable forms to use in numerical modeling of estuary flow.

The practical importance of vertical mixing is twofold. It determines the rate at which a pollutant introduced at a certain depth will spread throughout the depth of water, and, secondly, in conjunction with the current shear, it often plays a dominant role in horizontal dispersion. The time scale of vertical mixing may be expressed in terms of the time t during which an instantaneous discharge at a certain depth would spread so that its distribution in the vertical had a standard deviation σ_z, given by

$$\sigma_z{}^2 = 2K_z t. \qquad (5.21)$$

If σ_z is taken as 5 m, the time t would be about 35 hours for $K_z = 1$ cm^2 sec^{-1}, 3.5 hours for $K_z = 10$ cm^2 sec^{-1}, and 21 minutes for $K_z = 100$ cm^2 sec^{-1}.

These figures illustrate that in a strong tidal current, in homogeneous water, when K_z might reach several hundred cm^2 sec^{-1}, complete mixing in a depth of 10 m would occur in about half an hour. With K_z of the order of 10–20 cm^2 sec^{-1}, as in a typical weakly stratified estuary, such mixing would take an appreciable fraction of a tidal period, whereas in more strongly stratified cases it would take several days. In a layer with $K_z = 1$ cm^2 sec^{-1}, it would take over an hour for a discharge to attain a standard deviation of 1 m.

Whether rapid vertical mixing is an advantage or not depends on the objective in mind. To obtain a rapid reduction in the concentration of a discharged pollutant, good vertical mixing is an advantage. To achieve a speedy flushing of the pollutant from the estuary, on the other hand, it is an advantage if a discharge near the surface remains in the surface layer and is carried downstream by the net circulation without appreciable mixing into the lower layer.

The above approach to turbulent processes at middepths has been to start with the flow conditions in homogeneous water and to consider the modifications due to an increasing degree of vertical stratification. As the density gradient at middepth becomes greater, the density change becomes more and more concentrated in a thin layer, with relatively well-mixed layers above and below, extending to the surface and bottom, respectively. At this stage, the method of assuming continuous gradients of velocity and density and using coefficients of eddy viscosity and diffusion becomes less useful and may well be replaced by a two-layer representation, in which the exchange of momentum and material across the interface is represented by interfacial friction and exchange coefficients.

PARAMETERIZATION OF TURBULENT PROCESSES IN ESTUARY MODELS

In formulating a numerical model of circulation and mixing in an estuary, it is necessary to express the turbulent stresses and mass transports in terms of the mean velocity and the mean salinity or concentration of any other substance of interest. In a model of a partially mixed estuary in which the velocity and salinity are assumed to be continuous functions of the vertical coordinate, although represented in finite difference form, coefficients of eddy viscosity and diffusion have to be specified. One would expect the value of N_z, for example, to be proportional to the product of the rms turbulent velocity and a vertical mixing length. In homogeneous water, the rms turbulent velocity may be expected to be proportional to the mean current and the mixing length to the depth of water. In the stably stratified case, both the turbulent velocity and the mixing length may be expected to be reduced to an extent that is a function of the Richardson number.

A two-dimensional model of an estuary of simple geometrical shape was used by Bowden and Hamilton (1975) to investigate the effects on the circulation and mixing of changing various physical parameters. Changes in the velocity and salinity distribution during several tidal cycles were determined. Three forms of the coefficient of vertical eddy viscosity N_z and diffusion K_z were examined. In case (a) they were assumed constant throughout the tidal period; in (b) they were dependent on the depth–mean current and the depth of water, thus varying during the tidal period; and in (c) they were also functions of the overall Richardson number. In each case the coefficients were assumed to be constant over the depth. The actual values were as follows:

(a) $N_z = 40 \text{ cm}^2 \text{ sec}^{-1}, \qquad K_z = 20 \text{ cm}^2 \text{ sec}^{-1};$

(b) $N_z = 5 + 0.25H|U|, \qquad K_z = 2.5 + 0.125H|U|;$

(c) $N_z = 5 + 0.25H|U| (1 + 7 R_i)^{-1/4},$
        $K_z = 2.5 + 0.25H|U| (1 + R_i)^{-7/4};$

where H is the depth of water in meters, $|U|$ is the magnitude of the depth–mean current in cm sec^{-1}, and R_i is given by Eq. (5.20).

Figure 5.4 shows the velocity and salinity curves as functions of time during a tidal period for these three cases. The station considered is in a uniform part of the estuary with a width of 1 km and a mean depth of 16 m. In each case, the tidal amplitude at the mouth was 1 m and the river discharge and salinity at the mouth were held constant. The velocity curves have several features in common, such as the ebb lasting longer than the flood, typical of a tidal wave progressing into shallow water. The velocity decreases throughout the depth on the ebb, and on the flood it increases to a middepth maximum, features to be expected because of the density current effect. The salinity curves also have some features in common, but there are significant differences. When the eddy coefficients are variable instead of constant, the flood and ebb parts of both velocity and salinity curves become more unsymmetrical and vertical mixing is more intense on the flood than on the ebb. These features correspond to observations in several estuaries, including the Mersey, Severn, and Rotterdam Waterway.

Figure 5.5 shows the variation of the eddy coefficients during a tidal period. In case (b), when N_z and K_z vary with $H|U|$ but are independent of R_i, the coefficients fall to very low values at slack water, reach maximum values on the flood, and reach somewhat lower values on the ebb because of the distortion of the tidal wave. In case (c) the variation is enhanced, since the Richardson number becomes very large near slack water, and the values of N_z and K_z are reduced even further. They would, in fact, become negligibly small if the small constant terms in the equations for N_z and K_z were not included.

A rather similar investigation has been carried out by Blumberg (1975) in another two-dimensional, laterally averaged model, but he has allowed also for the variation of the coefficients N_z and K_z with depth. He took a different form for the function $\phi(R_i)$, expressing the dependence of the coefficients on the Richardson number, i.e.,

$$K_z = K_0 \, \phi(R_i), \qquad (5.22)$$

where K_0 is the value in homogeneous water and is itself a function of z. Figure 5.6, taken from Blumberg's paper, illustrates the form of $\phi(R_i)$ as proposed by different workers. The various forms agree in making K_z and also the ratio K_z/N_z decrease with increasing R_i, but the rates of decrease differ significantly.

These examples have been given to illustrate that the values assigned to coefficients such as those of vertical eddy viscosity and diffusion have an important effect on the results given by numerical models. The same comment applies to the specification of the shearing stress at the bed. At present, the forms of these parameters are usually chosen empirically to fit a particular set of data.

CONCLUSIONS

Further work is needed in several aspects of the investigation of turbulent processes in estuaries.

1. More field measurements are needed of the turbulent velocity fluctuations, their spectral distribution, and their relation to external parameters and the mean flow. These would provide a sounder physical basis for the study of estuarine processes and, by comparison with observations in the laboratory and in the atmospheric boundary layer, indicate to what extent knowledge of turbulent flows in these fields may be applied to estuaries.

2. The measurements should include direct determinations of the Reynolds stress in the bottom boundary layer

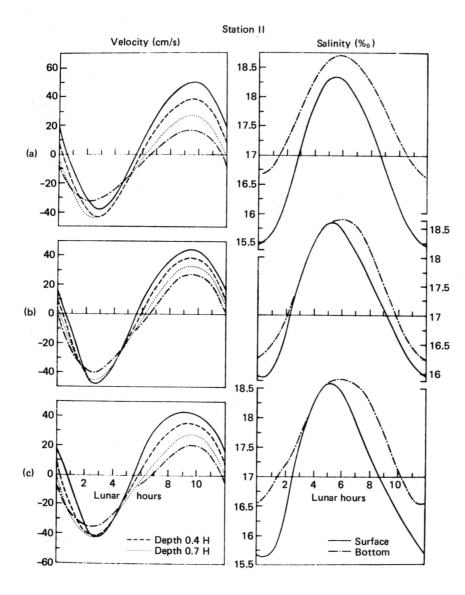

FIGURE 5.4   Velocity-time and salinity-time
curves in a model estuary for three forms of
the eddy coefficients N_z and K_z: (a) constant,
(b) variable, (c) dependent on Richardson
number R_i. From Bowden and Hamilton
(1975).

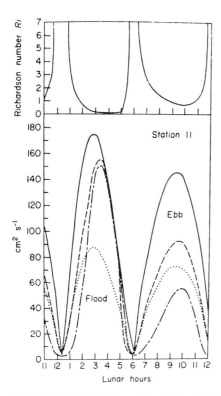

FIGURE 5.5 Variation of variable N_z (—) and K_z (. . . . .); and of Richardson-number-dependent N_z (- - -) and K_z (- · - · -) during the tidal cycle. From Bowden and Hamilton (1975).

by the eddy correlation technique and the relations of the stress to the turbulent intensity, $\overline{u'^2} + \overline{v'^2} + \overline{w'^2}$ or to $\overline{u'^2}$ alone. Simultaneous measurements should be made of the mean velocity profile near the bed so that the validity of the logarithmic law for determining the bottom stress could be better assessed.

3. Direct determinations are needed of the shear stresses and turbulent transports at middepths in relation to the stability, as represented by the Richardson number or some other suitable parameter. These would provide an improved basis for specifying the eddy coefficients that are needed in numerical models of circulation and mixing.

4. The time dependence of the above processes within the tidal cycle should be investigated, in view of the changes in the pattern of flow, the stability, and the turbulent intensities that are known to occur. Even if one's interest is in the longer-term effects, these cannot be properly assessed on the basis of mean values over a tidal period alone.

REFERENCES

Blumberg, A. F. (1975). A numerical investigation into the dynamics of estuarine circulation, Chesapeake Bay Inst. Tech. Rep. No. 91, The Johns Hopkins U., Baltimore, Md.

Bowden, K. F., and L. A. Fairbairn (1952). Further observations of the turbulent fluctuations in a tidal current, *Phil. Trans. R. Soc. A, 244,* 335.

Bowden, K. F., and L. A. Fairbairn (1956). Measurements of turbulent fluctuations and Reynolds stresses in a tidal current, *Proc. R. Soc. A, 237,* 422.

Bowden, K. F., and P. Hamilton (1975). Some experiments with a numerical model of circulation and mixing in a tidal estuary, *Estuarine Coastal Marine Sci. 3,* 281.

Bowden, K. F., and M. R. Howe (1963). Observations of turbulence in a tidal current, *J. Fluid Mech. 17,* 271.

Bradshaw, P., D. H. Ferris, and N. P. Atwell (1967). Calculation of boundary-layer development using the turbulent energy equation, *J. Fluid Mech. 28,* 593.

Fischer, H. B. (1972). Mass transport mechanisms in partially stratified estuaries, *J. Fluid Mech. 53,* 671.

Francis, J. R. D., H. Stommel, H. G. Farmer, and D. Parson (1953). Observations of turbulent mixing processes in a tidal estuary, Woods Hole Oceanographic Institution Rep. 53-22, Woods Hole, Mass.

Gordon, C. M. (1974). Intermittent momentum transport in a geophysical boundary layer, *Nature 248,* 392.

Gordon, C. M., and C. F. Dohne (1973). Some observations of turbulent flow in a tidal estuary, *J. Geophys. Res. 78,* 1971.

Grant, H. L., R. W. Stewart, and A. Moillet (1962). Turbulence spectra from a tidal channel, *J. Fluid Mech. 12,* 241.

Heathershaw, A. D. (1974). Bursting phenomena in the sea, *Nature 248,* 394.

McCave, I. N. (1973). Some boundary layer characteristics of tidal currents bearing sand in suspension, *Mem. Soc. R. Sci. Liege,* 6 ser., *6,* 107.

Mollo-Christenson, E. (1973). Intermittency in large-scale turbulent flows, *Ann. Rev. Fluid Mech. 5,* 101.

Phillips, O. M. (1967). The maintenance of Reynolds stress in turbulent shear flow, *J. Fluid Mech. 27,* 131.

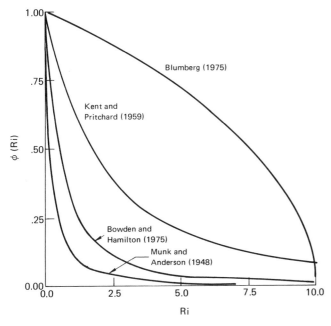

FIGURE 5.6 Various formulations of the stability function ($\phi(R_i)$) in $K_z = K_0\phi(R_i)$. From Blumberg (1975).

Pritchard, D. W. (1954). A study of the salt balance in a coastal plain estuary, *J. Marine Res. 13,* 133.

Pritchard, D. W. (1956). The dynamic structure of a coastal plain estuary, *J. Marine Res. 15,* 33.

Seitz, R. C. (1973). Observations of intermediate and small-scale water motion in a stratified tidal estuary (Parts I and II), Chesa-peake Bay Inst. Tech. Rep. No. 79, The Johns Hopkins U., Baltimore, Md.

Wiseman, W. J., R. M. Crosby, and D. W. Pritchard (1972). A three-dimensional current meter for estuarine applications, *J. Marine Res. 30,* 153.

The Coastal Boundary Layer

GABRIEL T. CSANADY
Woods Hole Oceanographic Institution

INTRODUCTION

Most direct human impact on the ocean occurs close to shore, within a zone where the effect of the coast on the hydrodynamics of the flow is pronounced. This zone is the "coastal boundary layer" (CBL), occupying the first 10 km of water in an offshore direction and having a depth of the order of 10 m. Several physical factors combine to make the CBL singular in its hydrodynamic behavior, a fact that has become apparent only recently, after sufficiently dense observations were carried out in nearshore waters. The practical importance of this recognition is considerable for domestic and industrial waste disposal, nearshore beach erosion and many other problems, as the flow structure in the CBL controls mass transfer rates and has other important influences.

The physical factors that combine to distinguish the CBL from other regions of the ocean are the following:

1. The shallowness of the water, which makes wind stress an overwhelming influence compared with any realistic longshore pressure gradient;

2. The offshore slope of the bottom, which has the dynamical effect of allowing long shore-trapped waves to be generated by intermittent storms;

3. The usual stable stratification of the water column, which insulates bottom layers from the direct influence of the wind stress and causes a concentration of longshore momentum in the top layers.

The effects of these physical factors are complicated by the Coriolis force due to the earth's rotation. They may be understood, however, in terms of a few simple conceptual models, such as the topographic wave, the coastal jet, and the internal Kelvin wave. The effects of friction are also, of course, relatively important and contribute further to the contrast between nearshore and offshore regions.

Especially important from a practical point of view is the resultant pattern of onshore-offshore flow, which governs mass transfer in an offshore direction. Current work is aimed at a quantitative determination of offshore transfer rates.

In the strict sense of the word, an estuary is the tidal mouth of a large river where fresh and saline waters meet and interact. In an extended sense, one may include in the concept that larger portion of the sea influenced materially by the freshwater–saltwater exchange. As Ketchum and Keen (1955) and Stommel and Leetmaa (1972) demonstrated, waters covering the continental shelf exhibit many estuarine traits of behavior, and their circulation is partly driven by the density difference between inflowing freshwater and the saline waters of the ocean.

In between estuaries in the strict sense, the continental shelf is bounded on the land side by more or less straight pieces of coastline. Seaward of the coastline lies a band of water within which oceanic water movements adjust to the presence of shores. This is the CBL, which has recently been found to exhibit a very different climatology of currents, temperatures, and salinities than waters further offshore. The discovery has considerable practical importance because, to use a currently fashionable phrase, most human "impact" on the ocean takes place via the CBL. Any effluents discharged nearshore are transported by the currents in the CBL and dispersed by their shear and turbulence. How this transport and dispersion works, and at what rate CBL waters exchange with offshore waters, largely determines to what extent such effluents become hazardous or objectionable. Pollution is, however, not the only important problem affected by CBL mechanics; beach movement and biological productivity of the continental shelves, to give two further examples, are also vitally influenced and are in the long run probably a great deal more important than effluent dispersal.

Self-evident as the practical importance of the CBL may be, what is surprising is that this band of the sea has not been subjected to serious physical investigations until recently. In the theoretical literature of physical oceanography, there have been indications from time to time that the coastal zone may well possess some singular properties, but there has been no systematic analysis of this question, much less a corresponding observational study, until a few years ago. Work on CBL dynamics got under way first in the Great Lakes, presumably because their store of freshwater makes these lakes an even more important local resource than the ocean. The CBL studies carried out, especially in Lake Ontario, have yielded results sufficiently general to be relevant to nearshore dynamical problems in many other locations. In this chapter, I shall try to present the more striking observed physical properties of the CBL and to relate these to conceptual models that appear useful in their understanding.

EMPIRICAL DEFINITION OF THE COASTAL BOUNDARY LAYER

The first clear distinction between current regimes observed near and far from shore has apparently been drawn by Verber (1966). In the course of a large-scale experiment conducted by the U.S. Public Health Service in Lake Michigan, a number of current meters were deployed, covering the whole lake in a more or less even grid pattern. Some of the

meters were near shore in shallow water. In classifying the types of current regimes observed, Verber noted that "straightline flow" is always found near shore, while far from shore the currents generally oscillate in all directions. Figures 6.1 and 6.2, taken from Verber's paper, illustrate

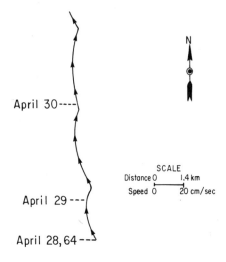

FIGURE 6.1 Currents observed at a station about 5 km from the western shore of Lake Michigan, April 1964, showing 2-h progressive vectors. Shoreline orientation is parallel to the mean current direction; the depth of water, 22 m. From Verber (1966).

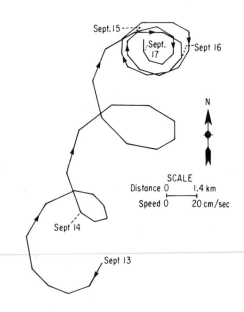

FIGURE 6.2 Currents at 20 km from the eastern shore of Lake Michigan, in water 60 m deep. Shoreline orientation is NNW to SSE. From Verber (1966).

the typical current behavior, by means of a progressive vector diagram, near and far from shore.

The same data were later analyzed in greater detail by Birchfield and Davidson (1967) and Malone (1968). Spectra at stations far from shore were dominated by a large peak at frequencies slightly above inertial, while close to shore most of the energy was in low-frequency motions. According to Malone (1968), there is a "large current component nearly parallel to the coast" at near-shore stations. The percentage of energy in the near-inertial peak is 3 times less near shore than offshore.

In addition to such characteristic differences between currents, large thermocline movements near shore were reported under stratified conditions. The periodic occurrence of upwelling of cold water near shore has been a well-known phenomenon around the Great Lakes for a long time. Investigations of the thermal structure of these lakes by Church (1942), Ayers *et al.* (1958), Mortimer (1963), and others also established that strong tilts of the thermocline upward or downward only occur in a nearshore band roughly 10 miles wide. Figure 6.3, from Mortimer (1963), shows a cross section of Lake Michigan with both shores subject to upwelling. This is somwhat unusual, because usually one shore exhibits a downwelled thermocline while the other is upwelled, but it illustrates the point that thermocline slopes are much steeper near shore than elsewhere.

When strong horizontal temperature contrasts are present, the nearshore zone usually comes to be separated from the offshore region by a thermal front (Tikhomirov, 1963; Rodgers, 1965). The persistence of fronts of this kind implies the presence of strong nearshore currents (Ragotzkie,

1966). Figure 6.4, from Ragotzkie (1966), illustrates the Keewenaw current in Lake Superior, a coastal current associated with a nearshore thermal front.

All this early evidence on the distinct character of the coastal zone became consolidated and placed into perspective by work during and in preparation for the International Field Year on the Great Lakes (IFYGL), carried out in 1972 on Lake Ontario. By the time the planning of this cooperative experiment got under way, there was considerable evidence, observational and theoretical, to suggest the unique importance of the first 10 km or so from the coast. Accordingly, detailed coastal zone observations were incorporated into the Field Year objectives and were in fact carried out for several years prior to the Field Year in Lake Ontario, along both the north and south shores (Csanady, 1972; Scott and Landsberg, 1969). Figures 6.5 and 6.6, from Csanady (1972), illustrate distinct coastal currents associated with a thermocline upwelling and downwelling, respectively. These were obtained on occasion of daily surveys using instruments lowered overboard from a small boat.

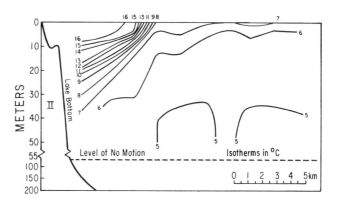

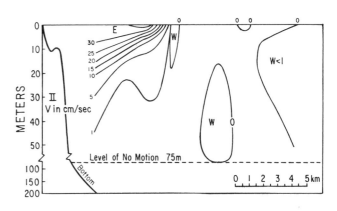

FIGURE 6.4 Temperature distribution (°C top) and inferred longshore geostrophic velocities (cm sec^{-1}, bottom) in Keewenaw current in Lake Superior, about halfway along north shore in Keewenaw Peninsula, July 1965. Illustration shows cross section perpendicular to shore. From Ragotzkie (1966).

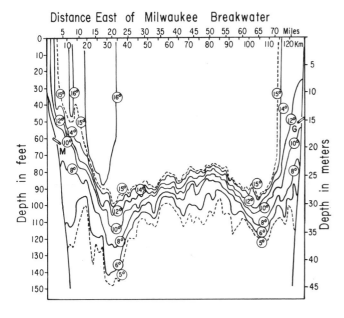

FIGURE 6.3 Temperature distribution on a Lake Michigan cross section, Milwaukee to Grand Haven. Note sharp isotherm slopes near shore. From Mortimer (1963).

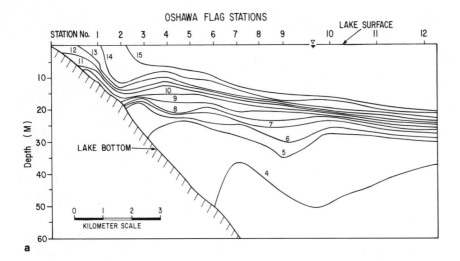

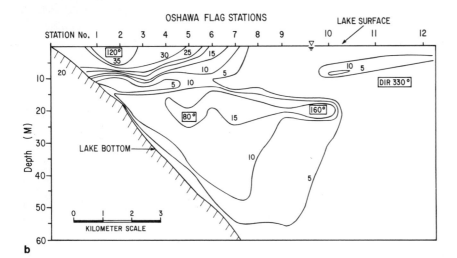

FIGURE 6.5 (a) Isotherms, °C and (b) constant-speed contours, cm sec^{-1}, in a CBL cross section on the north shore of Lake Ontario, during upwelling episode on September 29, 1970. Direction of flow is essentially long-shore and eastward (80°). From Csanady (1972).

Moored-current-meter observations in the same period provided further evidence. Weiler (1968) has shown that the kinetic energy of surface currents peaks some 8 km from shore. Blanton (1974) has corroborated this finding and showed that the percentage of energy in long-period motions falls dramatically between 8 and 10 km from shore. Blanton (1975) has also demonstrated that the typical onshore–offshore component of the surface currents near shore is much smaller than the longshore current, typically by a factor of 5.

Various other details of flow within the CBL have also been elucidated during and following the IFYGL, but the main point should now be clear; water motions within the first 10 km or so of the coast differ qualitatively from those offshore. In particular, they are characterized by relatively intense longshore currents.

The prevalence of longshore flow and the weakness of onshore–offshore velocities has important consequences for the nearshore dispersal of effluents (Csanady, 1970). Effluent plumes tend to hug the shores and to remain trapped in the shore zone for long distances. Figure 6.7 illustrates

Rhodamine B plumes generated by dyeing the cooling water outflow of a power station on Lake Huron. Most of the mass exchange between CBL and offshore waters seems to take place on those occasions when the coastal current reverses direction. Reversals take place roughly with the frequency of weather cycles, typically every 4 days; but there are other phenomena involved, more complex than direct driving by wind, as carefully demonstrated, for example, by Blanton (1975).

It remains to relate the evidence collected in the Great Lakes to open oceanic coasts. We have carried out some studies along the straight south coast of Long Island recently, and Figure 6.8 illustrates some of our results. These results have not been published before and are at present quite tentative. However, it seems that the following points may be made safely:

1. A clear distinction may be made between a nearshore zone (i.e., a coastal boundary layer) of some 8-km width and waters further offshore.
2. The tidal velocity amplitude is of order 20 cm sec^{-1}

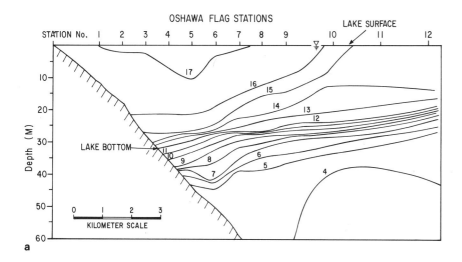

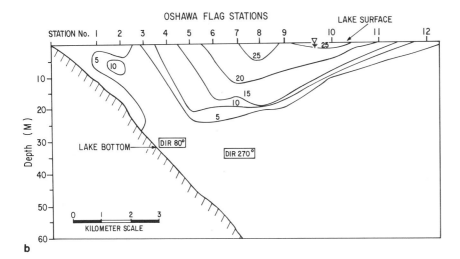

FIGURE 6.6 The same nomenclature as used in Figure 6.5 but for downwelling episode on September 17, 1970. Direction of higher speed flow is westward. From Csanady (1972).

and has a moderately strong influence on observed currents everywhere.

3. Within the CBL the nontidal currents are more responsive to winds and therefore more variable than further offshore.

4. The nontidal circulation further offshore (the midshelf circulation) is driven mostly by large-scale pressure gradients and shows relatively little variability (Scott and Csanady, 1976; Beardsley *et al.*, 1975).

5. Upward and downward tilts of the thermocline are present in the CBL, but they are noticeably less steep than in the Great Lakes. Intense mixing associated with tidal currents is believed responsible for the difference.

The remark that the nontidal currents within the CBL off Long Island are more variable than outside of it should be qualified; they are similar to currents within the CBL of the Great Lakes in that they respond fairly directly to winds. The contrast arises with the midshelf circulation, which is much steadier than midlake motions. The latter are domi-

nated by inertial oscillations and not by a steady pressure-gradient-driven flow, as the shelf circulation apparently is.

RELEVANT CONCEPTUAL MODELS

Water movements in large lakes or over continental shelves are exceedingly complex, containing (if resolved into some kind of spectrum) contributions from many widely different time and space scales. Full understanding of the details of these motions is impossible in principle because the human mined is able to comtemplate only a finite and in fact quite limited number of parameters at one time. Our only choice is a piecemeal understanding, obtained through "conceptual models," or idealized lakes or oceans of simple characteristics, subject to a limited number of external influences. If a distinct observable phenomenon in a real lake or ocean is reproduced by a simple conceptual model, we may be said to understand that phenomenon, in the sense that we know what factors cause it and what makes it stronger or weaker.

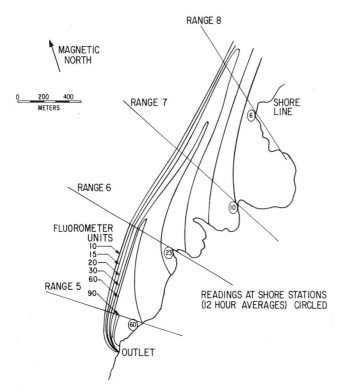

FIGURE 6.7 Contours of constant dye concentration in cooling-water plume leaving Douglas Point power station on Lake Huron, summer 1966, with coastal current heading northward. Plume remained within 1 or 2 km of shore for its entire detectable length, of the order of 10 km. From Csanady (1970).

Several conceptual models are necessary to understand the phenomena observed in CBL's. Some of these are known today, some remain to be discovered. A few conceptual models known to be relevant to CBL mechanics are briefly described below. They give among them a fairly well-rounded picture of why the CBL behaves as it does, even if they are presumably far from complete.

1. *Topographic gyres.* Consider a closed basin of realistic topography, its depth being of order 100 m near the center and reducing gradually to zero near the shores.* Let a constant wind stress pointing in a specific direction (e.g., along the longest axis of the basin) be applied at the free surface at time $t = 0$. Assume that the water is homogeneous, Coriolis force due to the earth's rotation negligible, and friction absent. The question is, "What motions are produced in the basin by the wind?" The linearized tidal equations with a wind-forcing term adequately describe the dynamical response for some initial period.

* A continental shelf may be thought of as a similar basin with one shore missing. Nearshore phenomena are not seriously affected by the presence or absence of an opposite shore, given only that the basin is wide enough for CBL's to exist.

Solutions of this problem may be found analytically for simple basin shapes and numerically for quite arbitrary topography. Figure 6.9 illustrates the response of Lake Ontario to an eastward stress, as calculated numerically by Rao and Murty (1969). The figure shows streamlines of transport (depth-integrated velocity) and is characterized by gryes related to the depth distribution, which are called topographic gyres.

The detailed physical reasons for the occurrence of topographic gyres are easily elucidated (Bennett, 1974; Csanady, 1975). A pressure gradient is set up by the wind stress, exactly balancing the latter in water of average depth. In shallower water, the wind stress overwhelms the pressure gradient and the flow accelerates downwind. In deeper water, the pressure gradient dominates and a return flow develops. Also, near shore the transport has to be accommodated in a shallow column, resulting in high velocities. The net result is fast downwind flow in the CBL, where the water is only of order 10 m deep, versus depths of the order of 100 m further offshore. Return flow is slow and tends to be lost in the background noise of other water motions. The fast, wind-driven coastal currents stand out and are the principal reason for the distinct character of the CBL.

2. *Topographic waves.* Observed coastal currents do not always point downwind. Especially following a sharp wind-stress impulse, flow within the CBL may well be found to oppose the direction of the wind stress, which evidently set up the flow field in the first place. Such phenomena may be explained by invoking the conceptual model of a topographic wave (Csanady, 1976a).

Over a sloping beach, any offshore or onshore displacements of fluid columns lead to stretching or squashing of vortex lines on a rotating earth. The implied generation of relative vorticity acts as a restoring torque in the vorticity equation and allows the existence of a vorticity wave related to the topography of the sea floor—a "topographic wave." The vorticity initially generated near shore in setting up the topographic gyre propagates as a topographic wave along the basin perimeter in a cyclonic (counterclockwise, in the northern hemisphere) direction. This is very much a shore-trapped wave in the sense that its amplitude decays rapidly with distance from shore. Given the topography of Lake Ontario, the wave is imperceptible beyond about 15 km from shore. Its speed of propagation is of the order of 50 cm sec^{-1}.

The detailed mechanics of topographic wave development are as follows. The strong coastal currents established in a topographic gyre adjust to geostrophic balance within a fraction of an inertial period. On the right-hand shore (looking along the wind that set up the gyre), this implies a slight rise of water level, and on the left-hand shore a slight drop. Thus a longshore pressure gradient comes into existence, mainly across the upwind and downwind ends of the basin. The water accelerates down the gradient, with the result that the stagnation points of the flow, and with them the entire pattern of coastal currents, rotate counterclockwise.

Figure 6.10 illustrates an observed instance of nearshore

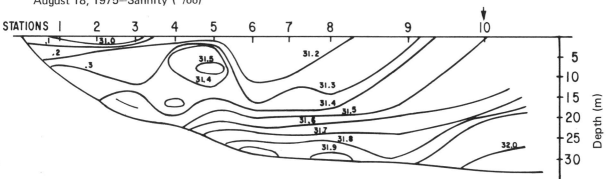

FIGURE 6.8 Longshore velocity (cm sec^{-1}, broken lines eastward, full lines westward), density (σ_t), temperature (°C), and salinity (‰). Contours in cross section of coastal boundary layer south of Long Island, near Shinnecock Inlet. The first eight stations are spaced 1 km apart. August 18, 1975, case of downwelled pycnocline and nearly uniform westward flow.

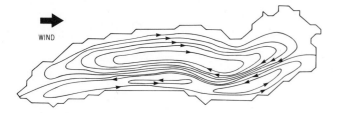

FIGURE 6.9 Calculated transport distribution in Lake Ontario with eastward wind stress driving. From Rao and Murty (1969).

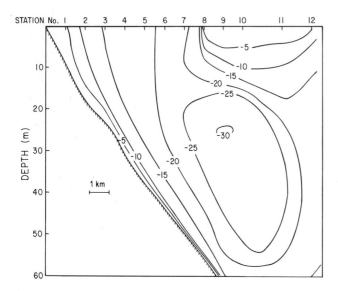

FIGURE 6.10 Longshore velocity distribution along north shore of Lake Ontario several days after cessation of an eastward wind stress impulse. Negative velocities shown in this figure are westward, opposite to the direction of the last storm. From Csanady (1976a).

flow, which could be shown to be principally due to a topographic wave passing this section, surveyed some 4 days after the generation of the wave by a storm. The extra surface elevation accompanying waves of this kind is a few centimeters, or barely detectable. The longshore velocities, on the other hand, are strong, 30 cm sec^{-1}, and contribute importantly to local current climatology in the CBL.

3. *Coastal jets.* Under stratified conditions, any external influence (such as a wind-stress impulse) usually generates motions also in an "internal" mode. An internal mode motion implies the rearrangement of constant density surfaces, e.g., an up or down movement of the thermocline, coupled to opposite offshore and onshore movements above and below, respectively. The equations describing internal mode motions are of a boundary-layer character (Csanady, 1972, 1976b), meaning that the highest space derivatives are multiplied by a small factor. That factor is the squared ratio of the internal radius of deformation to basin width.

The typical magnitude of the internal radius of deformation is 5 km, as against basin width at least 10 times larger, in basins of the size of Lake Ontario. Internal mode motions therefore are often very different in a baroclinic boundary layer of scale width 5-km than further offshore.

Specifically, wind-stress impulses cause baroclinic coastal jets to develop in the CBL. Within these jets, in a band of the order of 5 km width, all the wind-imparted momentum remains confined to the top layer above the thermocline. This means that the motions generated are considerably faster than in more homogeneous currents. The internal geostrophic balance of these motions also requires large thermocline slopes, often creating a thermocline outcropping on the free surface (a fully upwelled thermocline) or its opposite—a completely downwelled thermocline. Figure 6.11 is a schematic illustration of coastal jets in an idealized canal.

Figure 6.12 shows an observed coastal jet accompanied by a fully upwelled thermocline in a relatively pure form, more or less immediately after a wind-stress impulse leaving the coast to the *left*. An opposite wind stress, leaving the coast to the right, produces a thermocline downwelling. A change from one thermocline structure to another involves a massive exchange of water between the CBL and the offshore region.

4. *Kelvin waves.* The thermocline elevations and depressions accompanying coastal jets are subject to effects rather similar to those discussed under topographic waves. Looking along the wind that sets up the coastal jets, the right-hand shore comes to develop a downwelling, the left-hand shore an upwelling. Internal pressure gradients thus become established in a longshore direction across the upwind and downwind ends of the basin. These are again confined to a nearshore band, much as in a topographic wave, although the scale width (the internal radius of deformation) is even less. The resulting shore-trapped wave is known as an internal Kelvin wave, and its speed of propagation is also typically close to 50 cm sec^{-1}, the sense counterclockwise.

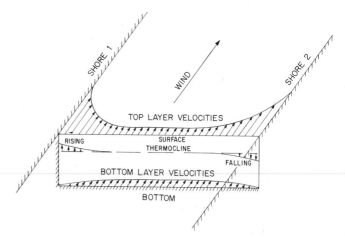

FIGURE 6.11 Schematic illustration of coastal jet conceptual model in two-layer flow. Csanady (1976b).

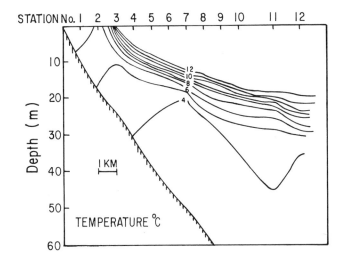

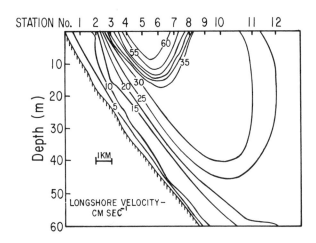

FIGURE 6.12 Temperature (top, °C) and longshore velocity (bottom, cm sec^{-1}) distribution along north shore of Lake Ontario immediately after eastward storm. Velocities are all eastward. From Csanady (1976a).

As a Kelvin wave passes a given location on shore, the direction of the coastal jet reverses, the thermocline tilt switches from upwelling to downwelling or vice versa. Some spectacular examples of such current and thermocline reversals have been observed during the IFYGL, and one case is shown here in Figures 6.13 and 6.14. The location of the observations is shown on the map of Figure 6.15.

5. *Frictional effects.* That a lateral boundary should cause a frictional boundary layer to form is evident from classical fluid mechanics. In a rotating, wind-driven basin, lateral boundary layers must in addition accept fluxes from interior Ekman layers, as discussed in detail by Greenspan (1968). Birchfield (1972, 1973) has pointed out some interesting complications arising from an interaction of such friction-rotation effects with bottom slope in a coastal zone, yielding in effect another mechanism for producing a CBL

scaled by frictional influences. Unfortunately, our knowledge of turbulent friction is at present too rudimentary to tie these theoretical results conclusively to observation. However, if we parameterize friction in the conventional way, using an eddy viscosity, the CBL comes to be proportional to the fourth root of the Ekman number times the basin width. In Lake Ontario, a realistic estimate of this kind of CBL width is perhaps 15 km.

CONCLUSION

The experimental evidence seems to be conclusive in support of a distinct CBL current climatology and density structure. Several physical factors are seen to cooperate in producing marked differences in behavior between nearshore and offshore currents. The most important application of the various conceptual models would be to estimate mass exchange rates between CBL and offshore waters. This depends on the pattern of onshore–offshore velocities, which are almost an order of magnitude smaller than the longshore flow and therefore relatively hard to observe accurately.

Two main problems face us in the task of parameterizing onshore–offshore mass exchange. One is the conceptual one of quantifying the contribution to this exchange of the various phenomena involved, e.g., current reversals consequent upon the passage of a topographic or Kelvin wave. The other problem is observational, to determine with sufficient accuracy and resolution the onshore velocity component within and just outside the CBL. The new generation of current meters with linear sensors (electromagnetic and acoustic), which is now coming into wide use, appears capable of delivering the required information. Work is in progress on these tasks, but support so far is not sufficient to accomplish the work that needs to be done.

ACKNOWLEDGMENT

This work has been supported by the Brookhaven National Laboratory under its ERDA-sponsored coastal shelf transport and diffusion program and by the Great Lakes Environmental Research Laboratory of NOAA.

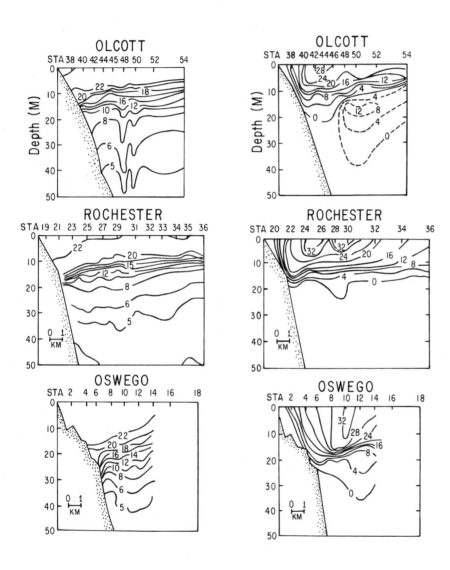

FIGURE 6.13 Isotherms (left) and constant longshore velocity contours (right) along south shore of Lake Ontario at three coastal chains on July 23, 1972, after a day-long eastward storm. Velocities shown by full lines point eastward. Downwind coastal jets are clearly present at all three stations, all accompanied by thermocline downwelling. From Csanady and Scott (1974).

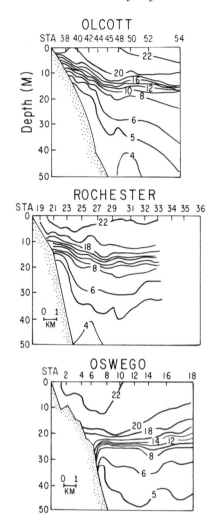

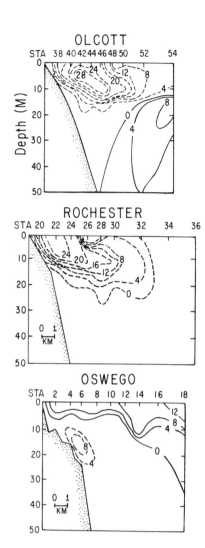

FIGURE 6.14 Isotherms and constant velocities along south shore of Lake Ontario on July 31, 1972. The eastward storm began on July 22, lasted till July 27, and was followed by four days of quiescent weather. Broken lines indicate reversed (westward) velocities, which appeared first at Olcott (already during the eastward storm), then at Rochester. The Oswego section is on the point of reversal. Note that the thermocline tilt has also reversed at Olcott and Rochester, from a downwelling to upwelling. From Csanady and Scott (1974).

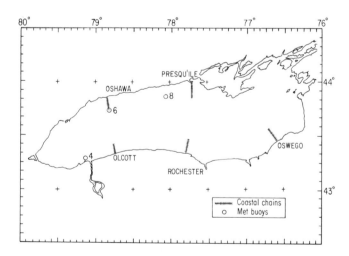

FIGURE 6.15 Locations of coastal observations ("coastal chains") in Lake Ontario during the IFYGL. From Csanady and Scott (1974).

REFERENCES

Ayers, J. C., D. C. Chandler, G. H. Lauff, C. F. Powers, and E. B. Henson (1958). Currents and water masses of Lake Michigan, U. of Mich., Great Lakes Research Inst. Publ. No. 3, Ann Arbor, Mich.

Beardsley, R. C., W. C. Boicourt, and D. V. Hansen (1975). Physical Oceanography of the Middle Atlantic Bight (to be published).

Bennett, J. R. (1974). On the dynamics of wind-driven lake currents, *J. Phys. Oceanog. 4*, 400.

Blanton, J. O. (1974). Some characteristics of nearshore currents along the north shore of Lake Ontario, *J. Phys. Oceanog. 4*, 415.

Blanton, J. O. (1975). Nearshore lake currents measured during upwelling and downwelling of the thermocline in Lake Ontario, *J. Phys. Oceanog. 5*, 111.

Birchfield, G. E. (1972). Theoretical aspects of wind-driven currents in a sea or lake of variable depth with no horizontal mixing, *J. Phys. Oceanog. 2*, 355.

Birchfield, G. E. (1973). An Ekman model of coastal currents in a lake or shallow sea, *J. Phys. Oceanog. 3*, 419.

Birchfield, G. E., and D. R. Davidson (1967). A case study of coastal currents in Lake Michigan, *Proceedings of the Tenth Conference on Great Lakes Research,* U. of Mich., Ann Arbor, Mich., pp. 264-273.

Church, P. E. (1942). Annual temperature cycle of Lake Michigan, U. of Chicago Dept. of Meteorology Misc. Rep. 18, Chicago, Ill.

Csanady, G. T. (1970). Dispersal of effluents in the Great Lakes, *Water Res. 4*, 79.

Csanady, G. T. (1972). The coastal boundary layer in Lake Ontario, *J. Phys. Oceanog. 2*, 41, 168.

Csanady, G. T. (1975). Hydrodynamics of large lakes, *Ann. Rev. Fluid Mech. 7*, 357.

Csanady, G. T. (1976a). Topographic waves in Lake Ontario, *J. Phys. Oceanog. 6*, 93.

Csanady, G. T. (1976b). The coastal jet conceptual model in the dynamics of shallow seas, in *The Sea,* Vol. VI, in press.

Csanady, G. T., and J. T. Scott (1974). Baroclinic coastal jets in Lake Ontario during IFYGL, *J. Phys. Oceanog. 4*, 524.

Greenspan, H. P. (1968). *The Theory of Rotating Fluids,* Cambridge U. P., Cambridge, England.

Ketchum, B. H., and D. J. Keen (1955). The accumulation of river water over the continental shelf between Cape Cod and Chesapeake Bay, *Deep Sea Res. Suppl. 3,* 346.

Malone, F. D. (1968). An analysis of current measurements in Lake Michigan, *J. Geophys. Res. 73*, 7065.

Mortimer, C. H. (1963). Frontiers in physical limnology with particular reference to long waves in rotating basins, U. of Mich., Great Lakes Res. Inst. Publ. No. 10, Ann Arbor, Mich.

Ragotzkie, R. A. (1966). The Keweenaw current, a regular feature of summer circulation of Lake Superior, U. of Wisc. Dept. of Meteorology Rep., Madison, Wisc.

Rao, D. B., and T. S. Murty (1969). Calculation of the steady-state wind driven circulation in Lake Ontario, *Arch. Meteorol. Geophys. Bioklimatol. A, 19,* 195.

Rodgers, G. K. (1965). The thermal bar in the Laurentian Great Lakes, *Proceedings of the Eighth Conference on Great Lakes Research,* U. of Mich., Ann Arbor, Mich., pp. 358-363.

Scott, J. T., and D. R. Landsberg (1969). July currents near the south shore of Lake Ontario, *Proceedings of the Twelfth Conference on Great Lakes Research,* U. of Mich., Ann Arbor, Mich., pp. 705-722.

Scott, J. T., and G. T. Csanady (1976). The COBALT experiment: first current meter observations, unpublished.

Stommel, H., and A. Leetmaa (1972). The circulation on the continental shelf, *Proc. Nat. Acad. Sci. U.S. 69,* 3380.

Tikhomirov, A. I. (1963). The thermal bar of Lake Ladoga, *Izv. Akad. Nauk SSR Ser. Geograf. 95,* 134. (Translation: Am. Geophys. Union Soviet Hydrology Selected Papers No. 2.)

Verber, J. L. (1966). Inertial currents in the Great Lakes, *Proceedings of the Ninth Conference on Great Lakes Research,* U. of Mich., Ann Arbor, Mich., pp. 375-379.

Weiler, H. S. (1968). Current measurements in Lake Ontario in 1967, *Proceedings of Eleventh Conference on Great Lakes Research,* U. of Mich., Ann Arbor, Mich., pp. 500-511.

II

ENGINEERING AND ENVIRONMENTAL INVESTIGATIONS

Water-Quality Analyses
of Estuarine Systems

7

DONALD J. O'CONNOR, ROBERT V. THOMANN, and DOMINIC M. DI TORO
Manhattan College

INTRODUCTION

The purpose of this chapter is to indicate the use of various analytic and mathematical methods in addressing water-quality analysis of estuarine systems and specifically to demonstrate how these analyses were used to project water-quality conditions for a variety of geophysical conditions and remedial measures. The ability to project future conditions in a reliable and realistic manner is one of the essential elements on which environmental planning of any natural water system is based. The ability to make such projections rests equally on the fundamental understanding of these systems as well as on a realistic application of these principles to specific problems. The chapters in Part I address the former point, the scientific component, while this and the remaining chapters in Part II are directed to the latter, its engineering and environmental counterpart.

This chapter includes a brief description of the general principles that are used in structuring estuarine models and a brief description of the validation procedure and of the application of the analysis for projection and planning. This is followed by a series of examples of water-quality analyses

in various estuarine systems throughout the country. The concluding remarks relate to the general validity and utility of the present state of the art of water-quality analysis and some general recommendations concerning the development and direction of these models.

GENERAL DESCRIPTION OF ANALYSIS

Mathematical analyses of water quality in estuarine systems, as in any natural water system, are based on quantitative relationships between inputs to and the water-quality response of these systems. The common basis of the majority of these analyses is the principle of continuity or mass balance. This principle is not only of fundamental importance in that it provides a means of structuring realistic models and evaluating their validity but is of practical value in that it also provides a means for projecting future water quality for a variety of geophysical conditions and remedial alternatives. Given a water-quality problem in a specific area, the particular constituents that relate to that problem are defined. A mass balance is developed for one or more of these interacting constituents, which takes into account three factors: the transport through the system, the reactions within it, and the inputs into it. The first factor describes the hydrologic and hydrodynamic regime of the water system; the second, the biological, chemical, and physical reactions that affect the water-quality constituents; and the third, the inputs or withdrawals of the substances through man's activities and natural phenomena.

Each estuarine system has its own geomorphological structure and hydrometeorological regime that establishes the transport component. Municipal, industrial, agricultural, and natural inputs of mass affect the concentration of various constituents that may create a water-quality problem. Identification and definition of the particular problem and the related constituents lead to a specification of the reactions that are relevant to that problem. These relationships have been in a state of continuous development over the past few decades; for many of the pertinent water-quality parameters, they are reasonably well formulated at present. Depending on the geomorphological and hydrological characteristics of the water system, the transport may be incorporated in a similar fashion, i.e., in a relatively simple or complex manner. The degree of simplicity or complexity is predicated on the temporal and spatial variation of the particular substance that is under consideration. The simplest framework is steady-state one-dimensional space. In the complex case, the use of hydrodynamic relationships may be required to quantify the transport terms.

The inputs are next considered in the formulation of the equations. The inputs of mass to the water system are due to the discharges from municipal, industrial, and agricultural activities and the runoff from these areas and from the natural drainage of undistributed regions. The inputs of various constituents from municipal sources are well known, with respect to both the average values and their variations. This is also true of many industries that are characterized by the production of one or a few products—the pulp and paper, canning, and steel industries, for example. However, industries that produce a variety of products, such as organics, synthetic chemicals, and pharmaceuticals, are more difficult to characterize. The temporal variability of these inputs with respect to that of the transport terms is an important factor in the structuring of the equation. Each area thus has a set of specific characteristics that qualitatively describe the problem and that may be expressed quantitatively as transport, input, and reaction coefficients. The transition from general qualitative principles to a specific quantitative formulation is effected by assigning a set of realistic coefficients. That such coefficients are realistic is determined by the degree to which the concentration, as computed by the model, agrees with observed data for various flow, temperature, and input conditions.

The general principle of the conservation of mass, as described above, is used to formulate the equations of the various constituents of water quality. In their simplest kinetic form, the equations describe the distribution of conservative substances, such as dissolved solids, or single reactants, such as bacteria or virus. They increase in com-

plexity as consecutive reactions (dissolved oxygen) are addressed. In their most complex form, they incorporate the interaction of a number of constituents (eutrophication and the transport of toxic substances in aquatic food chains). Also critical is the delineation between point sources, which are readily controllable, and the nonpoint sources, which are relatively difficult and, in some cases, impossible to control. Water-quality responses due to each of these sources must be clearly distinguished and described. The difficulty of assigning realistic values to distributed nonpoint sources is evident, given the present state of knowledge and data. Their relative significance becomes increasingly important as higher levels of point-source treatment are enforced.

The general equations must now be structured to fit the conditions of a specific estuary. This step invariably involves a segmentation of the particular water body, taking into account the transport, reactions, and inputs. It should allow for a realistic portrayal of the advective and dispersive components of the transport. It should permit a reasonable representation of the kinetics, particularly with respect to geophysical and hydraulic characteristics of the system. Finally, it should provide for a description of the waste inputs, and tributaries, both point and distributed. The guidelines for segmentation are also dictated by the method of solution, i.e., analytically integrated forms or finite-difference techniques. Given the method of solution, the system is segmented such that the basic assumptions or postulates are maintained without violating certain mathematical or numerical requirements or introducing unnecessary computational complexity.

The process of validation involves comparision between computed values of a variable and those measured in the prototype. When this comparison is satisfactory, the model is said to be validated. What constitutes "satisfactory" depends on the nature of the problem, the stucture of the model, and the extent of available data.

The first step is the calibration of the model. Given the geophysical and hydrological characteristics of the system, and a water-quality constituent whose reactions are known in principle, an estimate is made of the appropriate coefficients and inputs. These may be known from auxiliary models, as in the case of hydrodynamic transport terms; they may be assigned from correlations developed in laboratory or field experiments, which is the usual case for the kinetic terms; or they may be directly measured, as in the case of inputs. Transport and kinetic coefficients may also be available from direct prototype observations and measurements. In any case, assuming that a range of these values is known, a best estimate is made of each, the model is run, and the output compared with the data. Invariably successive adjustments are required to obtain a "reasonable" agreement between the model and the data. After the model has been thus calibrated, it is then validated by repeating the procedure for a different set of flow, temperature, and input data. The repetition of this procedure for other combinations effects a greater degree of validation with each favorable comparison between computation and observation.

The validated models are then used to project water-quality

conditions for various control procedures or policies. Water-quality models, on the whole, are planning and analytical tools rather then predictive and forecasting techniques. They are used primarily to examine the range of responses of the system that may occur under varying combinations of hydrological and meteorological conditions for the immediate and long-term development of the area. The validated model permits a range of alternates to be evaluated in a realistic and quantitative manner. These projections provide one of the many inputs that the administrator needs in order to make appropriate decisions for environmental planning. The significance of the water-quality problem and the consequences of the decision should be examined from both the environmental and economic points of view. The degree to which the analyses, provided by the models, affect the decisions should be commensurate with the level of validity of the models themselves.

APPLICATIONS OF ANALYSIS

The following examples describe water-quality problems in various estuarine systems throughout the country. They have been selected to demonstrate how the analysis of the type described has been used in evaluation, projection, and planning. Two general classes of water-quality problem are reviewed: those that are dominated predominantly by bacteria and produced depressions of dissolved oxygen by virtue of the biological oxidation of organic carbon and ammonia and those that are influenced mainly by the algae, which utilized inorganic nutrients, phosphorus, and nitrogen as their food sources. These two classes of water-quality problem have significant environmental effects and economic consequences, since the majority of pollution-abatement programs are concerned with the treatment and removal of these substances, i.e., organic carbon (biochemical oxygen demand, or BOD), ammonia, and nutrients.

DELAWARE ESTUARY

The management of the water quality of the Delaware Estuary has been concerned primarily with the alleviation of the depressed dissolved oxygen conditions that exist in the 30-mile region below Philadelphia. As shown in Figure 7.1, this region of the estuary received nearly one million pounds/day of carbonaceous BOD and one-half million pounds/day of nitrogenous BOD. This combined discharge produced dissolved oxygen concentrations of less than 1.0 mg/liter during summer low-flow conditions.

The causes of this depletion are the bacterial oxidation of the reduced carbonaceous and nitrogenous material discharged to the estuary. Carbonaceous organic matter is oxidized by many ubiquitous heterotrophic species of bacteria, whereas the oxidation of ammonia and nitrate nitrogen is accomplished by a specific group of aerobic autotropic bacteria. These differing pathways require that the kinetics for each reaction be included in the analysis. Figure 7.2

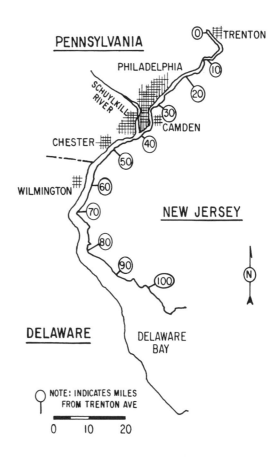

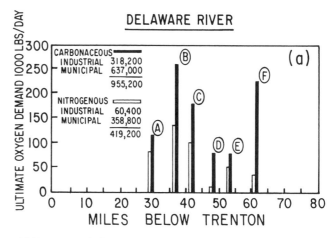

FIGURE 7.1 Delaware Estuary location map and BOD loadings. From O'Connor *et al.* (1968).

presents the results of such a calculation. The individual profiles refer to the individual sources in Figure 7.1. The oxidation of the nitrogenous component is inhibited by the low dissolved oxygen caused by the carbonaceous oxidation. Verifications of the model calculations are shown in Figure 7.3 for differing flow and temperature. Nitrogenous oxidation is also inhibited by low winter temperature.

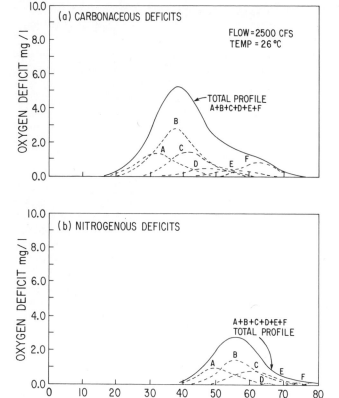

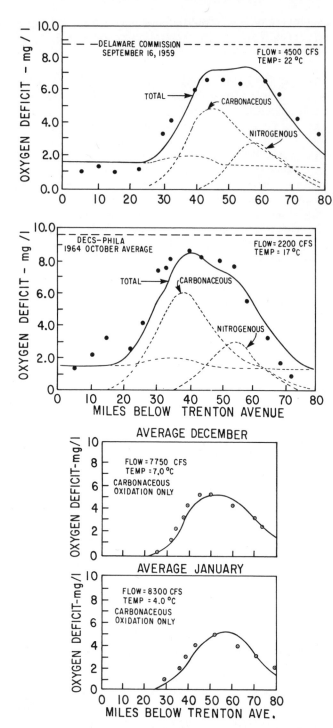

FIGURE 7.2 Calculated carbonaceous and nitrogenous deficits along the Delaware Estuary. From O'Connor *et al.* (1968).

The interplay of the two forms of oxygen demand complicate the projected response of the estuary to remedial control alternatives. As shown in Figure 7.4, reduction of 85 percent of the carbonaceous impact raises the minimum dissolved oxygen approximately 2 mg/liter, whereas the maximum deficit caused by this discharge approaches 5 mg/liter. The reason is that under projected conditions the dissolved oxygen concentration in the critical reach improves to the point where nitrogenous oxidation occurs in this region of the estuary rather than further downstream where a larger estuary volume is available to lessen the impact of the discharge.

The critical importance of the nitrogenous component of the oxygen demand prompted a more detailed examination. Instead of the simple reactant approximation, the individual steps of the nitrification reaction are analyzed: the hydrolysis of organic nitrogen compounds to ammonia, its partial oxidation to nitrite, and the subsequent complete oxidation to nitrate. Figure 7.5 illustrates a result of this analysis. The reaction coefficients required to achieve agreement are also indicated. The oxidation reactions are inhibited in the region of low dissolved oxygen as shown. The buildup of ammonia in this region is the result of the inhibited nitrification.

FIGURE 7.3 Observed and calculated oxygen deficit distributions for the Delaware Estuary. From O'Connor *et al.* (1968).

HOUSTON SHIP CHANNEL

The Houston ship channel is tributary to Galveston Bay, which in turn exits into the Gulf of Mexico. The tidal effects

DELAWARE RIVER

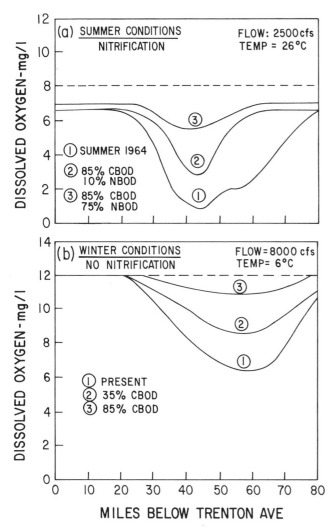

FIGURE 7.4 Projected dissolved oxygen distributions for the Delaware Estuary. From O'Connor *et al.* (1968).

are less and the extent of the depleted oxygen zone is greater. The increase in freshwater flow, Q, indicated in each figure, is due to the San Jacinto River. The range of the dissolved oxygen values reflects the difference between the surface and bottom concentrations. It is evident that this system is highly stratified. From a strictly scientific point of view, it might be argued that a two-dimensional analysis is necessary; however, from an engineering viewpoint, the vertically integrated model is sufficient for a preliminary analysis in order to assess the magnitude of the problem and as a screening mechanism to evaluate the various alternatives. As the range of alternatives is reduced, the two-dimensional analysis may be appropriate for greater refinement. The extended region of zero dissolved oxygen is caused by the high rate of waste inputs into a tidal system of low freshwater flow and minimal tidal exchange. Projections of water-quality conditions, based on secondary treatment of all the waste sources, are shown in Figure 7.8. Even with low flow augmentation of 500 and 1000 cubic feet/sec, low dissolved oxygen concentrations persist. The hydraulic influence of the augmented flow is relatively minor because of the large volume of water contained in the man-made channel and turning basin. This condition is characteristic of many estuaries throughout the country.

NEW YORK HARBOR

A further example of a tidal system is New York Harbor, a map of which is shown in Figure 7.9. The total mass discharge of BOD from municipalities and industries is much greater than that of the Houston ship channel. Dissolved oxygen (DO) concentrations, although depressed, are not so low as those in the channel. This is due primarily to a larger rate of tidal mixing and exchange and, to some degree, to the effect of the Hudson River freshwater flow. At the time the analysis was performed, the cross-hatched areas of New York City, as shown in Figure 7.9, were not receiving sewage treatment. The DO level has rarely dropped below about 1 mg/liter in the East River or below 2 mg/liter in the Hudson River, even when the waste discharge was greater than that shown in the figure. Since that time, treatment plants have been installed in these areas, except for the western part of Manhattan Island on the Hudson River, where a treatment plant is currently under construction. Figure 7.10 shows the comparison between the model and the data for dissolved oxygen in the Hudson River and the East River. The East River is well mixed vertically. Under low river-flow conditions, the Hudson is also; but as the freshwater flow increases, stratification becomes more pronounced and the low-DO harbor water intrudes upstream in the bottom layer of the Hudson. With projections of secondary treatment in the order of 65 to 85 percent removal of the significant waste sources, the DO may increase to about 4.5 mg/liter in the Hudson River and 3.5 mg/liter in the East River.

in the Bay are dampened by the restricted inlet, and hydraulic transport is further affected by the relatively low freshwater flows. Because of these characteristics, substances are retained in this system, particularly in the Houston ship channel, for extensive periods relative to those of the other estuaries studied here. Figure 7.6 portrays the distribution of freshwater flow and the location and magnitude of the wastewater inputs, as measured by the BOD, in the Houston ship channel. Morgan Point is at the junction of the Houston ship channel with Galveston Bay. Figure 7.7 presents the dissolved oxygen data and the calculated profiles for two periods. The range of the calculated profiles is based on a probable range of the nitrification rates of oxidation. At higher temperatures, the dissolved oxygen concentrations

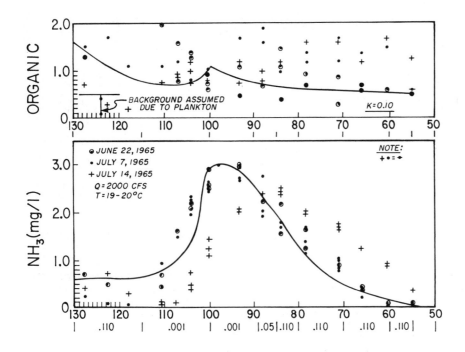

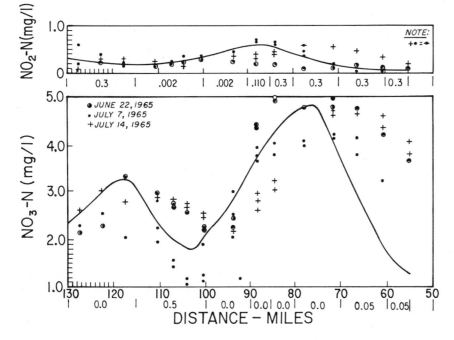

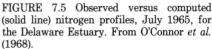

FIGURE 7.5 Observed versus computed (solid line) nitrogen profiles, July 1965, for the Delaware Estuary. From O'Connor *et al.* (1968).

BOSTON HARBOR

Tidal bays and harbors are frequently characterized by lateral as well as longitudinal variation of water quality. An example is the Boston Harbor area shown in Figure 7.11. The problem was the bacterial pollution that prevailed throughout the system, which necessitated closing the bathing beaches and the shellfish areas. The major sources of pollution were the effluents from two treatment plants at Nut Island and Deer Island, the storm overflows from the Charles River,

and the discharge of sludge from the treatment plants. Figure 7.12 presents the 1967 calculated and observed distribution of coliform bacteria, which indicated fairly good agreement. In the interval between 1967 and 1969, the construction of the Deer Island plant was completed and chlorination was put into operation. In 1969, then, the system received the chlorinated treated effluent from Deer Island, the storm overflows from the Charles River, and the sludge discharge from the plants. Figure 7.13 shows the 1969 coliform distribution. The calculated concentration

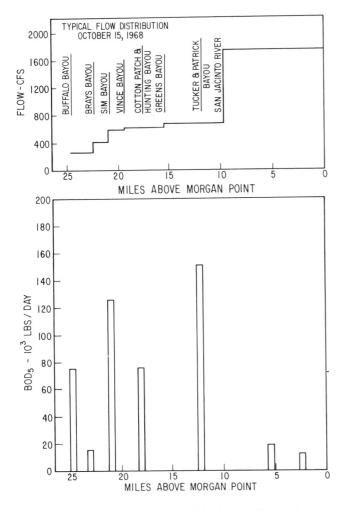

FIGURE 7.6 Spatial distribution of freshwater flow and waste-water inputs into the Houston ship channel. From Mulligan and O'Connor (1970).

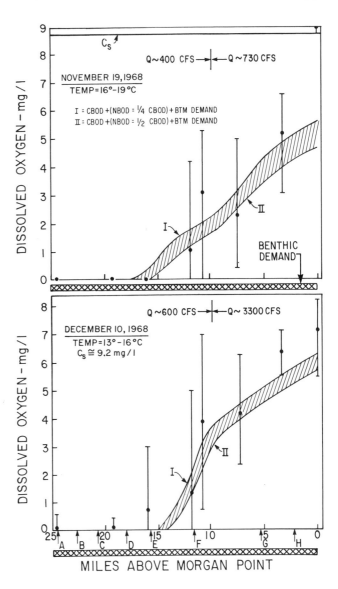

FIGURE 7.7 Dissolved oxygen profiles for the Houston ship channel. From Mulligan and O'Connor (1970).

south of Deer Island can be interpolated to be about 2000 MPN (most probable number) per 100 ml by contrast to the 120,000 MPN measured in 1967. The coliform contribution from the Nut Island and Deer Island effluents is now negligible, and it is apparent that the sludge disposal and the storm overflows are at present the significant sources of bacterial pollution. The relative importance of various inputs has been quantified in this example and indicates that alternatives other than advanced levels of treatment of point sources may result in greater improvement in water quality.

EUTROPHICATION ANALYSES

The final examples relate to the process of eutrophication of natural water systems, which is brought about by the discharge of nutrients, producing excessive amounts of

phytoplankton or aquatic weeds. The analysis of the problem is straightforward in principle. Inorganic nutrients, nitrogen and phosphorus, are converted to organic material through photosynthetic action. The plants usually grow during the summer and spring, when temperature and light conditions are conducive, and die in late fall, with little or no growth during the winter. They are also preyed upon by the next link in the food chain—the herbivorous zooplankton, and they, in turn, by the carnivores. These are then a food source for the next trophic level and, sequentially, through the food chain to man. Each trophic level by death and excretion returns organic nutrients, which hydrolyze to the inorganic form, and the cycle is maintained. Controlled discharges of nutrients to natural water systems are incorporated in this natural cycle and may be helpful and beneficial to

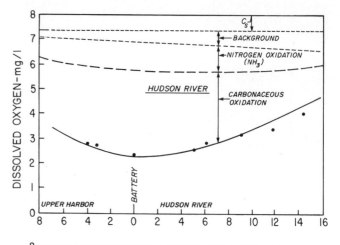

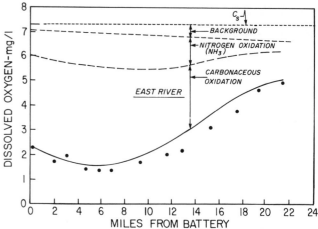

FIGURE 7.8 Projected dissolved oxygen profiles with flow augmentation for the Houston ship channel. From Mulligan and O'Connor (1970).

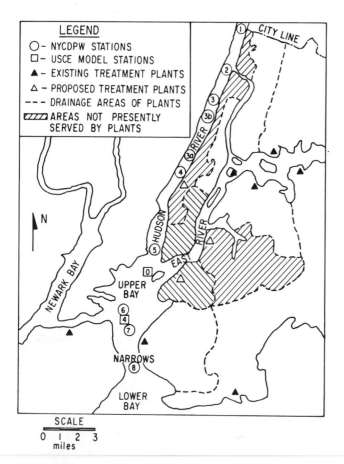

FIGURE 7.9 Map of New York Harbor. From O'Connor (1962).

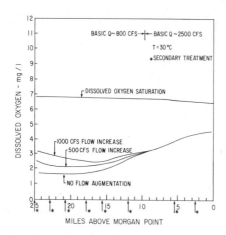

FIGURE 7.10 Dissolved oxygen distributions for the Hudson and East Rivers. From O'Connor (1962).

flow through the system. Reducing the hydraulic transport retains nutrients in the system for greater periods and thus increases their availability to phytoplankton.

SACRAMENTO-SAN JOAQUIN DELTA

It has been proposed to divert water from the Sacramento River in the north to the relatively arid areas in the south by means of a peripheral canal on the eastern border of the Sacramento-San Joaquin Delta, as shown in Figure 7.14. The question arose as to what effect the reduced flow through the delta and the downstream Suisun Bay would have on the potential eutrophication of this system.

Two specific geographical areas were analyzed in order to demonstrate the effect of freshwater flow—one at Mossdale, on the freshwater portion of the San Joaquin, and the second at Antioch, in the estuarine portion of the system. The annual variation of temperature, flow, and radiation for the years 1966 and 1967 is presented in Figure 7.15 for the San Joaquin River. These, in conjunction with the nutrient

productivity. It is the excessive discharge of nutrients that causes the degradation of water quality.

In addition to the availability of nutrients, other conditions affect the growth of phytoplankton, one of which is the

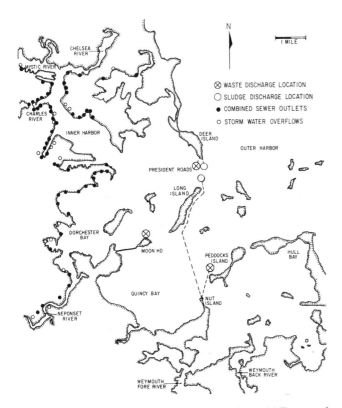

FIGURE 7.11　Area map of Boston Harbor. From Di Toro *et al.* (1973).

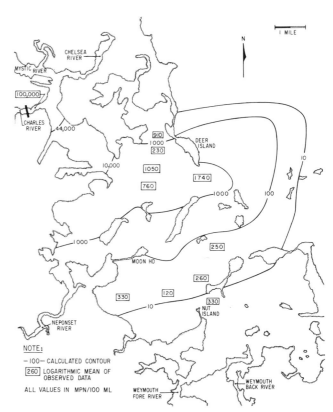

FIGURE 7.13　Total coliform verification, summer 1969, for Boston Harbor. From Di Toro *et al.* (1973).

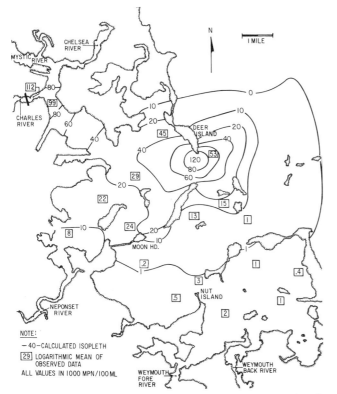

FIGURE 7.12　Total coliform verification, summer 1967, for Boston Harbor. From Di Toro *et al.* (1973).

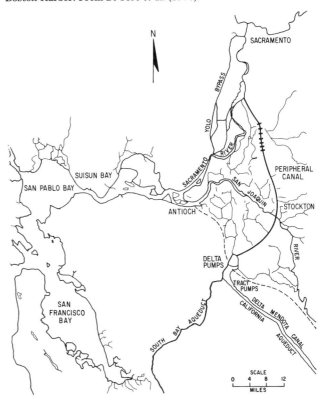

FIGURE 7.14　Location map for proposed peripheral canal project, Sacramento-San Joaquin Delta. From Di Toro *et al.* (1971).

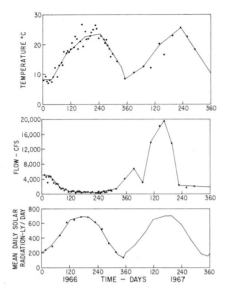

FIGURE 7.15 Temperature, river flow, and mean solar radiation for San Joaquin River at Mossdale. From Di Toro *et al.* (1971).

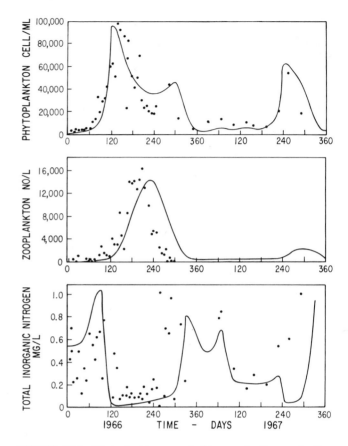

FIGURE 7.16 Phytoplankton, zooplankton, and nitrogen in the San Joaquin River at Mossdale. From Di Toro *et al.* (1971).

discharges, were input data for the model that calculated the temporal distribution of phytoplankton, zooplankton, and nutrients. These distributions are presented in Figures 7.16 and 7.17 for the freshwater San Joaquin River at Mossdale and for the estuarine waters at Antioch, respectively. The effect of the freshwater flow is evident; note the pronounced bloom in the spring of the low-flow year in 1966 by contrast to high flow in 1967, which essentially flushes the system before it has an opportunity to grow. The lower flow at the end of the year permits growth to take place, as evidenced by the fall bloom in 1967. In the estuarine area, the tidal mixing predominates over the freshwater flow and growth of phytoplankton occurs to about the same degree each year. The effect of freshwater flow in this region is obviously less pronounced.

Although there are differences between the calculated profiles and the observations, the general pattern is reproduced by the model, at least to the degree that permitted some preliminary evaluation of the effects of flow diversion and increased discharge of nutrients. The model was run for these projected conditions, and the results showed that increased nutrient inputs and increased light levels had a more pronounced effect than decreased flow.

POTOMAC ESTUARY

The basic issue addressed by the water-quality model of eutrophication of the Potomac Estuary is the expected phytoplankton response due to nutrient reductions of the Washington, D.C., Blue Plains plant. A substantial capital and operating cost for nutrient removal coupled with uncertainty over expected response in phytoplankton provide the setting

for a preliminary eutrophication model. The Potomac Estuary is shown in Figure 7.18 with the physical segments used in the model.

The calibration of the model utilized data collected during 1968. A second data set collected during 1969 under somewhat different flow conditions was then used for verification. Typical results are shown in Figure 7.19. The model generally captured the major interactions between the inorganic nutrients and phytoplankton growth for about the upper 50-mile reach of the estuary and for the spring–fall conditions. Transient late winter and early fall blooms during 1969 were not duplicated by the model. In general, however, the phytoplankton model provided a reasonable approximation to 1969 conditions using a 1968 "tuning" of the model.

In order to illustrate the application of the preliminary model, a simulation was prepared using 1969 conditions but with a 90 percent reduction of nitrogen and phosphorus at the municipal waste inputs. Figure 7.20 shows the results of the simulation. As indicated, the 90 percent reduction in nitrogen and phosphorus input reduces peak levels in July; but over about a 10-20 mile stretch of the estuary, phytoplankton levels will still be above the 25-50 μg/liter level sought by the Environmental Protection Agency as part of its proposal for nutrient reduction in Washington,

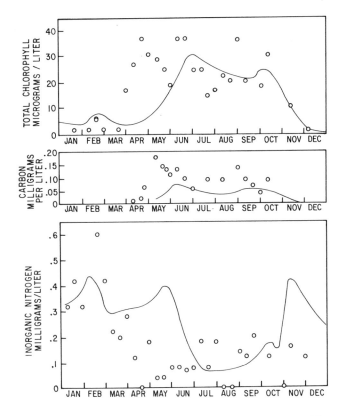

FIGURE 7.17 Phytoplankton, zooplankton, and nitrogen verification in San Joaquin Estuary at Antioch. From Di Toro *et al.* (1971).

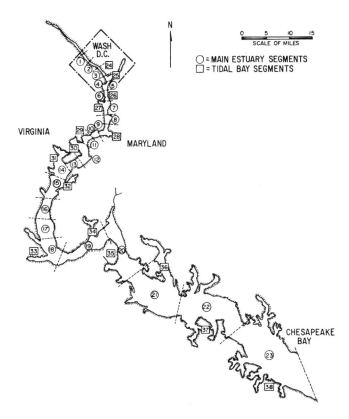

FIGURE 7.18 Map of Potomac Estuary showing longitudinal and lateral segments used in the model. From Thomann *et al.* (1974).

D.C. For a variety of reasons, including to some degree the results from this model, nitrogen removal at Blue Plains (representing a costly expenditure) has been delayed, and an intensive more detailed investigation has been initiated to determine the need for such removal.

CHESAPEAKE BAY

Increasing concern has been expressed in recent years over the apparent increase in evidence of cultural eutrophication in certain areas of the Chesapeake Bay system (Figure 7.21). For example, the approximate average phytoplankton chlorophyll in the upper bay in July 1950 was about 5 µg/liter, with maximum values of up to 25 µg/liter. In the period 1969–1971, however, average chlorophyll values were about 30 µg/liter, with maximum values exceeding 60 µg/liter. The interaction between nutrient discharges, specifically from the Baltimore area and the Susquehanna River, and the chlorophyll levels therefore formed an important link in determining nutrient allocations.

A simplified linear steady-state nutrient phytoplankton model was constructed to provide a basis for determining the relative effect of various nutrient inputs. The simplified linear model provided for rapid, relatively inexpensive com-

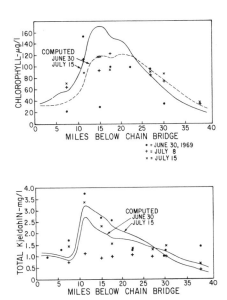

FIGURE 7.19 Spatial profile comparison of observed 1969 data and computed values for chlorophyll a (top) and Kjeldahl nitrogen (bottom) for the Potomac Estuary. From Thomann *et al.* (1974).

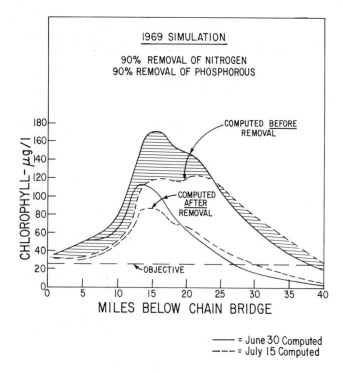

FIGURE 7.20 1969 simulation of chlorophyll for the Potomac Estuary. From Thomann *et al.* (1974).

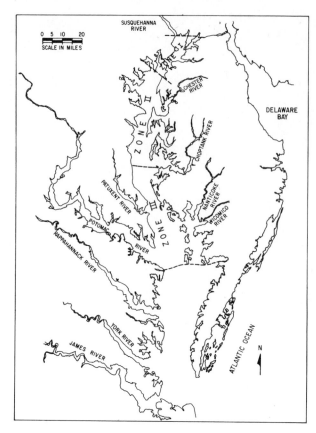

FIGURE 7.21 Geographical zones for projection analysis for Chesapeake Bay. From O'Connor *et al.* (1975).

putations for the first estimates of allowable nutrient discharges. The model is two dimensional in the lateral and longitudinal directions. Figures 7.22 and 7.23 show the comparisons between data and model output. Several sets of data were used for calibration and verification. As noted, the results indicate maximum average chlorophyll levels of some 40 μg/liter in the region between the entrance to the Susquehanna River and the Baltimore area. Figure 7.22 also shows that total phosphorus is not conserved in the water column, but losses occur probably due to settling of particulate forms. From the model analyses, it is estimated that during low river flows (5000 cfs), the Baltimore area and Susquehanna River phosphorus inputs account for 75 and 14 percent of the total, respectively. At high summer flows of 22,700 cfs, the Baltimore area and Susquehanna River phosphorus inputs are 50 and 43 percent, respectively. The model thus provided a first estimate of the relative importance of each source of nutrient and pointed toward the need for detailed examination of the interaction between the Susquehanna River and Baltimore Harbor effects.

The interactions were then modeled. The model permitted the decision-making function the opportunity to explore various trade-offs between phosphorus allocations, Susquehanna River flow and inputs, and chlorophyll water-quality objectives. The results of this modeling exercise provided input into a policy of phosphorus reduction and allocation by the State of Maryland.

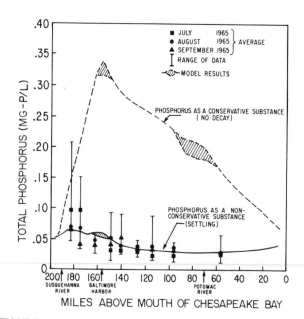

FIGURE 7.22 Model verification, summer 1965, for total phosphorus for Chesapeake Bay. From O'Connor *et al.* (1975).

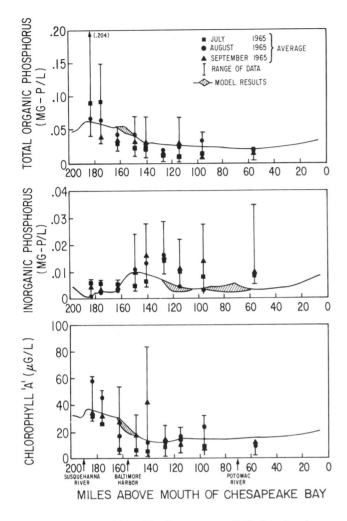

FIGURE 7.23 Model verification, summer 1965, for phosphorus and chlorophyll a for Chesapeake Bay. From O'Connor *et al.* (1975).

CONCLUSIONS

These examples have served to illustrate the input that estuarine quality models can provide to the decision-making process. In each case, the determination of the possible water-quality consequences of various environmental control actions was of some importance in the overall problem context. The ultimate decisions, of course, involve a variety of political, social, and economic issues; nevertheless, the need for describing the changes in water quality under different levels of external control is a key aspect of the overall process. Mathematical models, therefore, of estuarine water quality of certain classes of problems (bacteria, dissolved oxygen, and some aspects of eutrophication) have been developed, calibrated, and verified to a degree sufficient to permit credible use in the decision-making process.

REFERENCES

Di Toro, D. M., D. J. O'Connor, and R. V. Thomann (1971). A dynamic model of the phytoplankton population in the Sacramento-San Joaquin Delta, *Advan. Chem. 106*, 131.

Di Toro, D. M., J. Mancini, J. Mueller, D. J. O'Connor, and R. V. Thomann (1973). Development of hydrodynamic and time variable water quality models of Boston Harbor, Hydroscience Inc. report, Westwood, N. J.

Mulligan, T., and D. J. O'Connor (1970). Mathematical model and water quality analysis—Houston ship channel—Galveston Bay study, Hydroscience Inc. report, Westwood, N.J.

O'Connor, D. J. (1962). Organic pollution of New York Harbor—Theoretical considerations, *J. Water Pollut. Control Fed. 34*, 905.

O'Connor, D. J., J. P. St. John, and D. M. Di Toro (1968). Water quality analysis of the Delaware River Estuary, *Am. Soc. Civil Eng. J. Sanitary Eng. Div. 94*, SA6, 1225.

O'Connor, D. J., H. Salas, and R. V. Thomann (1975). The Chesapeake Bay waste load allocation study, Hydroscience Inc. report, Westwood, N. J.

Thomann, R. V., D. M. Di Toro, and D. J. O'Connor (1974). Preliminary model of Potomac Estuary phytoplankton, *Am. Soc. Civil Eng. J. Environ. Eng. Div. 100*, EE3, 699.

Real-Time Models for Salinity and Water-Quality Analysis in Estuaries

8

DONALD R. F. HARLEMAN
Massachusetts Institute of Technology

INTRODUCTION

The improvement and control of water quality in a natural water body such as a river or estuary can be achieved by intelligent regulation of municipal and industrial waste discharges. Waste-treatment techniques by chemical and biological processes are highly developed; and while it is technically possible to approach "zero discharge" of wastes, in most cases it is neither necessary nor economically feasible to do so. The important engineering decisions in water-quality control relate to the determination of the level of waste treatment that is consistent with the multiple uses of natural water bodies. This implies the ability to forecast or predict the response of the river or estuary to future increases in investment in waste-treatment facilities.

The prediction, by mathematical or physical models of effects and benefits, in advance of the construction of a facility, is the essence of engineering. Physical models are of limited use in water-quality studies because of the difficulty of simulating biochemical reaction processes at reduced temporal and spatial scales. Within the last decade, engineers and planners have made increasing use of mathematical models for planning purposes. The objectives of this chapter are to present some recent developments and future research directions for mathematical models for water-quality control and to illustrate the important dynamic coupling between hydrodynamic transport processes and biochemical water-quality transformation processes.

STRUCTURE OF WATER-QUALITY MODELS

All mathematical models are approximations, in varying degrees, of the natural processes that they attempt to represent in a deterministic manner. In an aquatic ecosystem as complex as a river or an estuary, there are a large number of possible approximations within the conceptual framework of simulation models. Water-quality models contain a large number of "rate constants" and almost any model can be made to agree with field data by the familiar process of curve-fitting. However, this is no guarantee that a particular model will give valid results when used in a predictive mode. It is important that the user or developer of a water-quality model have the ability to analyze the structure of the model and to judge the validity of basic assumptions.

Various strategies can be identified for structuring mathematical models of aquatic ecosystems. Table 8.1 identifies three basic types of model.

TABLE 8.1 Models of Aquatic Ecosystems

Type of Model	Organizing Principle	Causal Pattern	Measure
Biodemographic models	Conservation of species or of genetic information	Life cycles	Numbers of individuals or of species
Bioenergetic models	Conservation of energy	Energy-flow circuits	Energy, power
Biogeochemical models	Conservation of mass	Element cycles	Mass of elemental matter

Fisheries management models usually belong to the class of *biodemographic* models, as they are concerned with particular species and the processes that affect their numbers, such as birth, death, harvesting, and competition. *Bioenergetic* models are concerned with energy-flow, energy-storage, and energy-dissipation processes simultaneously coupling ecological and environmental components.

Most water-quality engineering models fall into the category of *biogeochemical* models. They employ the principle of conservation of mass to determine the distributions of dissolved oxygen, nutrients, and biomass by the coupling of hydrodynamic transport and biochemical transformation processes. The remainder of the discussion will be concerned with *biogeochemical* models as applied to water-quality control in rivers and estuaries.

BIOGEOCHEMICAL WATER-QUALITY MODELS

The hydrodynamic aspects of an estuary are the transport processes, which include the advection, mixing, and dispersion of specific constituents in waste effluents. In addition, these constituents are subjected to various *transformation* or *reaction processes,* leading to their production, decay, or both and to the production (or decay) of new biochemical compounds through interaction with the receiving water. Transport processes are relatively independent, or at least insensitive, to the characteristics of the wastes introduced into a waterway. The transformation processes, on the other hand, depend on both the transport processes and the interaction or coupling of constituents of the total waste load.

The essential features of biogeochemical models may be illustrated by considering the one-dimensional formulation in which cross-sectional areas, velocities, and concentrations are functions of longitudinal distance, *x,* and time, *t.*

ADVECTIVE TRANSPORT PROCESSES

In general, the transport processes are unsteady; therefore, the advective terms that describe the flow field must be determined by simultaneous solution of the continuity and momentum equations. The governing equations for one-dimensional unsteady flow in a variable area channel are the continuity equation,

$$\frac{\partial A}{\partial t} + \frac{\partial Q}{\partial x} = q, \qquad (8.1)$$

and the longitudinal momentum equation,

$$\frac{\partial}{\partial t}(AU) + \frac{\partial}{\partial x}(QU)$$

$$= -gA\frac{\partial h}{\partial x} - g\frac{Q|Q|}{AC^2 R_h} - \frac{gAd_c}{\rho}\frac{\partial \rho}{\partial x}, \qquad (8.2)$$

where

x is the distance along longitudinal axis,
t is time,
h is the elevation of water surface with respect to a horizontal datum,
Q is the cross-sectional discharge,
q is the lateral inflow per unit length of channel,
U is the average cross-sectional velocity in the channel, Q/A,
g is the acceleration of gravity,
A is the cross-sectional area of channel,
C is the Chezy roughness coefficient,
R_h is the hydraulic radius of channel,
ρ is the density of water,
d_c is the distance from surface to centroid of the cross section.

The last term in Eq. (8.2) represents the effect of a longitudinal density gradient. This term is significant only within the salinity intrusion region of estuaries. Boundary conditions must be specified (either water surface elevation h or discharge Q) at the upstream and downstream sections of the river or estuary being modeled. The solution of Eqs.

(8.1) and (8.2) can be obtained numerically by means of finite-difference schemes as described by Harleman and Lee (1969). The solution requires the specification of initial conditions for h and Q and advances in time in accordance with the values of the time-varying boundary conditions.

CONSERVATION OF MASS

The basic components of biogeochemical water-quality models are statements of conservation of mass. Essentially the model consists of a sequence of conservation of mass equations, one for *each* water-quality constituent.

The one-dimensional, conservation of mass equation may be written in the following form:

$$\frac{\partial}{\partial t}(AC) + \frac{\partial}{\partial x}(QC)$$

$$= \frac{\partial}{\partial x}\left(AE_L\frac{\partial C}{\partial x}\right) + A\left(\frac{r_i}{\rho} + \frac{r_e}{\rho}\right), \qquad (8.3)$$

where C is the concentration of a water-quality constituent (averaged over the cross section); E_L is the longitudinal dispersion coefficient; r_i is the time rate of *internal* addition of mass of substance C per unit volume by transformation of reaction processes; and r_e is the time rate of *external* addition of mass of substance C per unit volume by addition of substance across the lateral, free surface, and bottom boundaries of the system.

The hydrodynamic variables Q or U, and h or A, obtained by solution of Eqs. (8.1) and (8.2) are basic inputs to the transport terms on the left-hand side of Eq. (8.3). The longitudinal mixing process is represented by the first term on the right-hand side; and the internal and external "sources" or "sinks" of the particular water-quality constituent, represented by C, are contained in the last term.

Many transformation processes are dependent on the local temperature T and/or salinity S of the water, and it is convenient to determine $T = f(x,t)$ and $S = f(x,t)$ as state variables defining the water environment. The quantity ρcT represents the concentration of heat per unit volume of water. Therefore with $C = \rho cT$, Eq. (8.3) can be written as a conservation of heat equation:

$$\frac{\partial}{\partial t}(AT) + \frac{\partial}{\partial x}(QT) = \frac{\partial}{\partial x}\left(AE_L\frac{\partial T}{\partial x}\right) + \frac{\phi_n b}{\rho c}, \quad (8.4)$$

where T is water temperature, ϕ_n is the time rate of net heat input per unit area of water surface, and ρc is density (specific heat of water).

In Eq. (8.4), the last term represents a source of the external type, i.e., the net rate at which heat is transferred across the water surface by the combined processes of long-wave and shortwave radiation, evaporation, and convection.

Examples of the simultaneous solution of Eqs. (8.1), (8.2), and (8.4) to find $T = f(x,t)$ using numerical techniques are

given by Harleman *et al.* (1972). In rivers and in the uniform density portion of estuaries, not including the salinity intrusion region, the longitudinal dispersion term in Eqs. (8.3) and (8.4) is of secondary importance. Values of the longitudinal dispersion coefficient in rivers and estuaries may be estimated on the basis of the work of Holley *et al.* (1970).

In estuaries where salinity intrusion occurs, longitudinal dispersion becomes an important factor in the overall mixing process. This is due to the gravitational circulation induced by the freshwater–seawater density difference. The conservation equation for salinity is given by

$$\frac{\partial}{\partial t}(AS) + \frac{\partial}{\partial x}(QS) = \frac{\partial}{\partial x}\left(AE_{LS}\frac{\partial S}{\partial x}\right), \qquad (8.5)$$

where S is the salinity averaged over the cross section and E_{LS} is the longitudinal dispersion coefficient in the salinity intrusion region.

Equation (8.5) is an example of a "conservative" conservation of mass equation; in general, there are no internal or external sources of salinity except at the ocean boundary. Thatcher and Harleman (1972a, 1972b) use numerical techniques to solve Eqs. (8.1), (8.2), and (8.5) to find $S = f(x,t)$. They use a dispersion relationship in which E_{LS} is a function of the local longitudinal salinity gradient and the degree of vertical mixing in the estuary. Applications to unsteady salinity distributions in the Hudson, Delaware, and Potomac Estuaries under time-varying freshwater inflows are given in the references cited.

TRANSFORMATION PROCESSES: CBOD-DO MODELS

Historically, water-quality models have emphasized the importance of dissolved oxygen (DO) as the primary indicator of water quality. The earliest and most elementary example of a transformation process, within the framework of biogeochemical models, is the concept of carbonaceous biochemical oxygen demand (CBOD). This class of models has its origin in the water-quality studies of Streeter and Phelps beginning about 1925. The coupling between the set of conservation of mass equations for CBOD and DO is illustrated by the following forms of Eq (8.3): conservation of CBOD,

$$\frac{\partial}{\partial t}(AL) + \frac{\partial}{\partial x}(QL) = \frac{\partial}{\partial x}\left(AE_L\frac{\partial L}{\partial x}\right) - K_1 AL, \qquad (8.6)$$

where L is the ultimate CBOD and K_1 is the CBOD decay coefficient; and conservation of DO,

$$\frac{\partial}{\partial t}[A(\text{DO})] + \frac{\partial}{\partial x}[Q(\text{DO})]$$

$$= \frac{\partial}{\partial x}\left[AE_L\frac{\partial(\text{DO})}{\partial x}\right] - K_1 AL + K_2 A[(\text{DO})_s - (\text{DO})], \quad (8.7)$$

where DO is the concentration of dissolved oxygen, DO_s is the saturation concentration of dissolved oxygen, and K_2 is the surface reaeration coefficient.

The elementary coupling arises from the fact that the solution of Eq. (8.6), $L = f(x,t)$, appears in Eq. (8.7) as the internal decay term for dissolved oxygen. Dailey and Harleman (1972) have developed numerical methods for simultaneous solution of Eqs. (8.1), (8.2), and (8.4)–(8.7) in estuary networks of one-dimensional channels.

The fundamental limitations of the CBOD–DO class of models are well known and will not be discussed in detail. It is sufficient to say that it is essentially impossible to aggregate biogeochemical transformation processes within the concept of CBOD as a primary water-quality constituent. In addition, dissolved oxygen is not a sufficient water-quality indicator in that it provides no information on the state of eutrophication and the possibility of algal blooms. The application of modern waste treatment technology demands decisions on the removal of inorganic nutrients in addition to the conventional removal of oxygen-demanding organic materials. The questions of *what* to remove and *how much* to remove are fundamental to intelligent design and investment decisions on the treatment of wastes prior to discharge into an adjacent river or estuary. These questions can only be answered by water-quality models that are capable of *predicting* the ecological response of the waterway to increased levels of treatment. It is clear that future research efforts in water-quality modeling and field data collection should be directed to the modeling of transformation processes within the *element cycles* of an aquatic ecosystem.

TRANSFORMATION PROCESSES: NUTRIENT MODELS

Current efforts in water-quality modeling are attempting to deal with problems of eutrophication by assuming that aquatic ecosystems are composed of coupled conservation of mass equations (Quinlan, 1975). A one-dimensional, real-time, biogeochemical model for an aerobic, nitrogen-limited estuarine ecosystem subject to domestic sewage loadings has been developed by Najarian and Harleman (1975). The proposed estuary water-quality model attempts to follow the path taken by nitrogenous nutrients (in their various forms) based on the present knowledge of element cycles in aquatic ecosystems. In contrast to the simple CBOD–DO model, the nitrogen model represents various forms of organic and inorganic nitrogen that are the potential sources of eutrophic activity.

Jaworski *et al.* (1971) show nitrogen to be the limiting nutrient in the Potomac Estuary. Harrison and Hobbie (1974) arrive at a similar conclusion in their studies on the Pamlico River Estuary. Since estuaries are usually nitrogen-limited environments, the study of nitrogen may be sufficient to generate the necessary information for phytoplankton bloom predictions. Dugdale *et al.* (1967, 1971) give yet another reason for the choice of the nitrogen cycle for the study of phytoplankton dynamics, as follows:

Measurements of population growth using nitrogen may, in fact, show less scatter than would those using carbon or phosphorus, because the latter two elements are not only structural components but are continuously turned over in the energetic processes of organisms. Theoretically, nitrogen should provide an inherently satisfactory and fundamental measure of productivity in ecosystems.

The dynamic estuary nitrogen cycle model consists of a closed matter-flow loop having seven storage variables and twelve transformations of the element nitrogen from one storage form to another, as shown in Figure 8.1. The chosen storage and transformation processes represent physical, chemical, and biological forms of nitrogen. The hypothesized structure of the model is sophisticated enough to simulate nitrogen-limited ecosystem dynamics; yet it is simple enough to be amenable to computations. The seven storage variables include ammonia-N, nitrite-N, nitrate-N, phytoplankton-N, zooplankton-N, particulate organic-N, and dissolved organic-N. The biochemical and ecological transformations include nitrification, uptake of inorganic nutrients by autotrophs, grazing of heterotrophs, ammonia regeneration by living cells, lysis and leakage of organic matter through cell walls, natural death of microorganisms, and ammonification. The transformation rates are functions of nutrient concentrations and available energy in the form of heat and light.

To explore the dynamics of the closed nitrogen element cycle, the proposed seven-variable model is applied to the chemostat system shown in Figure 8.2. The set of seven mass conservation equations for the chemostat, obtained from Eq. (8.3) (with $\partial C/\partial x = 0$) are listed below. The assumed expressions for each of the transformation rates are shown on the solid arrows in Figure 8.1. The dotted lines indicate the information transfer necessary to determine the rates of matter flows.

$$\frac{dN_1}{dt} = R_{41}N_4 + R_{51}N_5 + R_{71}N_7 - R_{12}N_1 - R_{14} \cdot$$

$$\frac{N_1 N_4}{K_1 + N_1} + (Q/V)(N_1{}^0 - N_1), \qquad (8.8)$$

$$\frac{dN_2}{dt} = R_{12}N_1 - R_{23}N_2 + (Q/V)(N_2{}^0 - N_2), \qquad (8.9)$$

$$\frac{dN_3}{dt} = R_{23}N_2 - R_{34}\frac{N_3 N_4}{K_3 + N_3} + (Q/V)(N_3{}^0 - N_3), \qquad (8.10)$$

$$\frac{dN_4}{dt} = R_{14}\frac{N_1 N_4}{K_1 + N_1} + R_{34}\frac{N_3 N_4}{K_3 + N_3} - R_{45}\frac{N_4 N_5}{K_4 + N_4}$$
$$- (R_{41} + R_{46} + R_{47})N_4 + (Q/V)(N_4{}^0 - N_4), \qquad (8.11)$$

$$\frac{dN_5}{dt} = R_{45}\frac{N_4 N_5}{K_4 + N_4} - (R_{56} + R_{51})N_5 + (Q/V) \cdot$$
$$(N_5{}^0 - N_5), \qquad (8.12)$$

$$\frac{dN_6}{dt} = R_{46}N_4 + R_{56}N_5 - R_{67}N_6 + (Q/V)$$
$$(N_6{}^0 - N_6), \qquad (8.13)$$

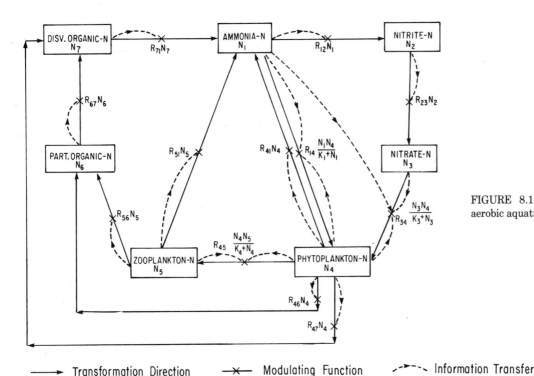

→ Transformation Direction   ✕ Modulating Function   ⌒‿ Information Transfer

FIGURE 8.1 Nitrogen cycle structure in aerobic aquatic ecosystems.

FIGURE 8.2 Batch and chemostat systems.

$Q = 0$ Boundaries closed to matter flow
$Q \neq 0$ Boundaries open to matter flow

$$\frac{dN_7}{dt} = R_{47}N_4 + R_{67}N_6 - R_{71}N_7 + (Q/V)$$
$$(N_7^0 - N_7), \quad (8.14)$$

where

N_1 is the concentration of NH_3-N, mg/liter;
N_2 is the concentration of NO_2-N, mg/liter;
N_3 is the concentration of NO_3-N, mg/liter;
N_4 is the concentration of phytoplankton-N, mg/liter;
N_5 is the concentration of zooplankton-N, mg/liter;
N_6 is the concentration of particulate organic-N, mg/liter;
N_7 is the concentration of dissolved organic-N, mg/liter;
N_i^0 is the concentration of ith nitrogen-cycle storage vari-

able in the influent discharge, mg/liter ($i = 1, 2, \dots 7$);
K_1 is the half-saturation constant for NH_3-N, mg/liter;
K_3 is the half-saturation constant for NO_3-N, mg/liter;
K_4 is the half-saturation constant for phytoplankton-N, mg/liter;
R_{ij} is the transformation rate from ith storage variable to jth storage variable, liters/day;
Q is the rate of inflow and/or outflow, ft³/day;
V is the volume of the container, ft³.

Equations (8.8)–(8.14) are applicable to batch and chemostat systems. However, the external source terms (Q/V) $(N_i^0 - N_i)$ are zero in batch systems. The analysis of the system equations reveals that the formulated structure of the nitrogen-cycle model is a closed-matter flow loop with no leaks. Indeed, for every positive internal source term there exists an identical negative internal sink term in the set of system equations. This reciprocity of the source/sink terms is in accord with the principle of mass conservation applied to a closed-loop element cycle.

The expressions that describe the rate of transformation processes are of two types: first-order reaction kinetics in the form of $dN_j/dt = R_{ij}N_i$, or saturation kinetics in the form of a hyperbolic transformation $dN_i/dt = R_{ij} [N_iN_j/(K_i + N_i)]$. Data in the literature show that transformation rates that relate to nutrient uptake by primary producers or predators usually follow saturation kinetics. These process rates are all temperature-dependent. Uptake rates also vary with the intensity of solar radiation. The rate governing parameters used in Eqs. (8.8)–(8.14) are given in Table 8.2.

TABLE 8.2   Transformation Rate Parameters

$R_{12} = 0.20$/day	
$R_{14} = 2.0$/day (light hours); 0.10 (dark hours)	
$R_{23} = 0.25$/day	
$R_{34} = 1.0$/day (light hours); 0.05 (dark hours)	
$R_{41} = 0.01$/day	
$R_{45} = 0.07$/day (light hours); 1.50 (dark hours)	
$R_{46} = 0.03$/day	
$R_{47} = 0.03$/day	
$R_{51} = 0.01$/day	
$R_{56} = 0.10$/day	$K_1 = 0.3$ ppm-N
$R_{67} = 0.30$/day	$K_3 = 0.7$ ppm-N
$R_{71} = 0.30$/day	$K_4 = 0.5$ ppm-N

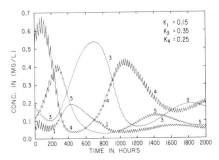

FIGURE 8.4   Dynamic response of a chemostat system with open boundaries. Curves: 1, NH_3-N; 3, NO_3-N; 4, Phytoplankton-N; 5, Zooplankton-N.

CHEMOSTAT SYSTEM

Figure 8.2 shows the chemostat system with mass transfer across the physical boundaries of the system. The mass conservation principle requires that the following constraint equation be satisfied:

$$\sum_{i=1}^{7} N_i(t) = \sum_{i=1}^{7} N_i^{in} + \int_0^t \sum_{i=1}^{7} [N_i^0 - N_i(t)] \frac{Q}{V} dt. \quad (8.15)$$

The response analysis of the nitrogen cycle in a chemostat system is shown in Figures 8.3 and 8.4. These analyses reveal some interesting system characteristics.

Figure 8.3 shows the dynamic response of the chemostat for certain initial values of the seven nitrogen storage quantities. The residence time is 310 hours (13 days), and it is observed that steady state is reached in about four residence times (approximately 1200 hours). Figure 8.4 shows the chemostat response for the initial conditions and residence time; however, the half-saturation concentrations K_1, K_3, and K_4 are reduced to 50 percent of the values assumed in Figure 8.3. All other rates governing parameters are the same in the two cases illustrated. The decrease of K_4 in the

simulation shown in Figure 8.4 has the effect of increasing the grazing rate of zooplankton on phytoplankton and results in higher zooplankton concentrations. The most significant effect of the reduction in half-saturation concentrations is the appearance of large damped oscillations of the nitrogen storage quantities. It is evident that steady state is not reached in six residence times (2000 hours).

IDEALIZED ESTUARY

The objective of this section is to investigate the coupling of the transport processes in an advective system (such as an estuary) with the biochemical nitrogen transformation processes. It is shown that the predicted concentrations of the nitrogen storage variables are highly sensitive to the representation of the real-time (i.e., intratidal cycle) tidal motion and dispersive mixing processes within the estuary. This is demonstrated by comparing model results for the real-time tidal motion with model results for the through-flow case. In the through-flow case, only the nontidal flow (due to freshwater inputs) is considered. This is equivalent to averaging the advective flows over a tidal period or, as some investigators have assumed, to considering only high or low water slack conditions in the estuary.

An idealized estuary having a uniform depth of 15 ft (4.6 m), length of 30,000 ft (9146 m), width of 1000 ft (305 m), and Manning roughness of 0.018 is assumed. A constant freshwater inflow rate of 1000 ft³/sec (28 m³/sec) enters the head of the estuary. The salinity at the ocean end of the estuary is 15,000 ppm, and the ocean tidal range is 4 ft (1.2 m). Two waste treatment plants are located on the estuary as shown in Figure 8.5. The flow rate of Plant I is 10 million gallons/day (mgd) (38,000 m³/day), and that of plant II is twice this value. The breakdown of the waste effluents in terms of the nitrogen storage variables is shown in Figure 8.5.

Transient flows in the estuary due to tide and freshwater inflows are calculated by numerical solution of the continuity equation (8.1) and the longitudinal momentum equation (8.2). In this application the temperature is assumed to be constant at 68 °F (20 °C).

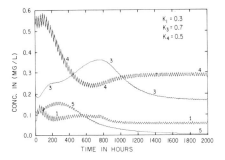

FIGURE 8.3   Dynamic response of a chemostat system with open boundaries. Curves: 1, NH_3-N; 3, NO_3-N; 4 Phytoplankton-N; 5, Zooplankton-N.

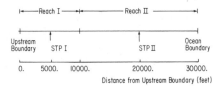

Description of Waste Injections

	STP I		STP II	
	Conc. (ppm)	Load (lb/day)	Conc.(ppm)	Load (lb/day)
CBOD	30	2502.	30	5004.
NH3-N	20	1668.	20	3336.
NO3-N	2	167.	2	334.
PON	10	834.	10	1668.
DON	10	834.	10	1668.

FIGURE 8.5 Schematic of ideal estuary and treatment plant loading.

The longitudinal salinity distributions and the x and t dependence of the longitudinal dispersion coefficient are calculated by numerical solution of Eq. (8.5). In addition, seven conservation of mass equations are needed for the seven nitrogen storage variables of the nitrogen-cycle model. Each of these equations is in the form of Eq. (8.3), in which the internal "source-sink" term r_i/ρ is replaced by the appropriate term from the right-hand side of Eqs. (8.8)–(8.14). It should be noted that the term $(Q/V)\,(N_i^0 - N_i)$ is to be omitted, since these are applicable only to the chemostat system. For the sake of brevity only the equation for ammonia (N_1) is given.

$$\frac{\partial}{\partial t}(AN_1) + \frac{\partial}{\partial x}(QN_1) = \frac{\partial}{\partial x}\left(AE_{LS}\frac{\partial N_1}{\partial x}\right) + AR_{41}N_4$$

$$+ AR_{51}N_5 + AR_{71}N_7 - AR_{12}\,N_1 - R_{14}\;\frac{N_1 N_4}{K_1 + N_1}. \qquad (8.16)$$

The numerical results from the real-time calculations for the idealized estuary are shown in Figures 8.6–8.9. Figure 8.6 shows the tidal discharge as a function of time throughout one tidal cycle at the ocean end and at section $x = 10,000$ ft (3049 m). The maximum tidal discharge at the ocean end is 9500 ft^3/sec (270 m^3/sec) and corresponds to a maximum tidal velocity of 0.65 ft/sec (0.2 m/sec). This may be compared with a freshwater through-flow velocity of 0.07 ft/sec (0.02 m/sec). This ratio of maximum tidal velocity to through-flow velocity is typical of many estuaries. Instantaneous longitudinal distributions of salinity in Reach II at four times during a tidal cycle are shown in Figure 8.7. The time $T/4$ corresponds to high-water slack and $3T/4$ to low-water slack. Since the longitudinal dispersion coefficient is assumed to be proportional to the local longitudinal salinity gradient, the dispersion coefficient increases significantly within the salinity intrusion region. In the nonsaline region, the dispersion coefficient is related to the local tidal velocity by a modified Taylor dispersion relation, which accounts for both vertical and lateral velocity distributions.

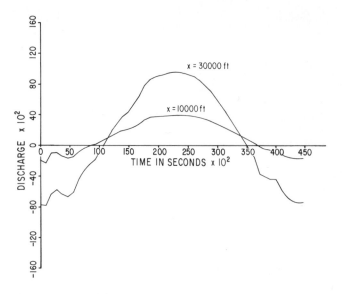

FIGURE 8.6   Tidal discharge versus time at $x = 10,000$ ft and $x = 30,000$ ft.

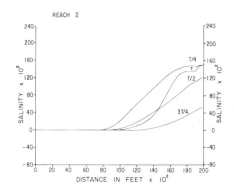

FIGURE 8.7   Longitudinal salinity profiles in Reach II.

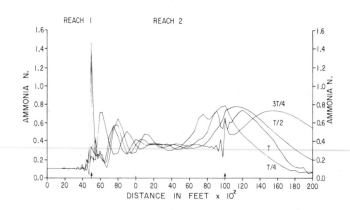

FIGURE 8.8   Ammonia concentrations in ideal estuary for real-time model.

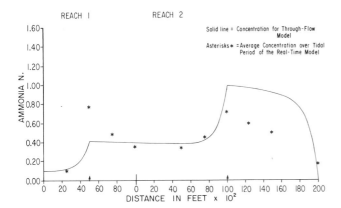

FIGURE 8.9 Ammonia concentrations in ideal estuary for through-flow model.

The complete dispersion relationship is given by

$$E_{LS(x,t)} = K \left| \frac{\partial S^\circ}{\partial x^\circ} \right| + mE_T, \qquad (8.17)$$

where $K = 50$ ft^2/sec (4.6 m^2/sec); $S^\circ = S/S_0$; S_0 is ocean salinity; $x^\circ = x/L$ and L is length of estuary. The Taylor dispersion coefficient, $E_T = 77\ unR_h^{5/6}$ (ft^2/sec), where u is the tidal velocity, n is the Manning roughness, R_h is the hydraulic radius, and $m = 3$ to account for the lateral velocity distribution. In the idealized estuary study, values of E_{LS} have averaged values of about 15 ft^2/sec (1.5 m^2/sec) in the nonsaline portion and reach maximum values of about 400 ft^2/sec (37 m^2/sec) in the salinity intrusion region.

Figure 8.8 shows instantaneous longitudinal profiles of ammonia (N_1) at four times in a tidal period. The large peaks of concentration adjacent to the upstream waste treatment plant in Reach I are due to the combined effect of low tidal velocities near the head of tide and low dispersion. The flushing effect near the ocean boundary is indicated by the large differences in ammonia concentration between low-water slack ($3T/4$) and high-water slack ($T/4$). Other constituents of the real-time nitrogen-cycle model exhibit similar longitudinal and temporal variations.

The idealized estuary computations were repeated using the nontidal, through-flow velocity due to freshwater inflow. A constant dispersion coefficient equal to the average of the spatially variable real-time coefficient (65 ft^2/sec) (6 m^2/sec) was used for this simulation. Figure 8.9 shows the predicted ammonia concentrations for the freshwater through-flow case. A comparison with the corresponding real-time simulation of Figure 8.8 is made by determining the average concentration over a tidal period and by plotting these averages in Figure 8.9. It is apparent that ammonia concentrations predicted by the through-flow model differ by as much as 100 percent from the real-time tidal averages even though the biochemical models are identical.

POTOMAC ESTUARY

The most dramatic evidence of the interaction of physical transport and biochemical transformation processes comes from the application of the nitrogen-cycle model to an actual estuary. As already demonstrated, the tidal motion and longitudinal mixing processes in an estuary are highly variable in both time and space; in addition, the diurnal light–dark cycle is out of plase with the lunar tidal cycle. Najarian and Harleman (1975) have demonstrated the application of the nitrogen-cycle model to the Potomac Estuary from the head of tide near Chain Bridge to the junction with Chesapeake Bay, a distance of 114 miles (184 km). A period of 19 tidal cycles (about 10 days) was chosen for the simulation run. Waste-treatment plant loadings in the upper 30 miles of the estuary were included, the major contribution being the Blue Plains waste-treatment facility. Time steps of the order of 30 min were used for the calculations, with spatial increments ranging from 1000 (305 m) to 10,000 ft (3050 m) in length.

Time histories of five of the nitrogen variables during the 10-day run at a location 2 miles below the Blue Plains waste-treatment facility are shown in Figures 8.10 and 8.11. Of interest is the large variability of many of the water-quality parameters within a tidal period.

TWO-DIMENSIONAL MODELS

The state of the art of two-dimensional, real-time models for salinity and water-quality analysis in estuaries is considerably less developed than that of the one-dimensional models. There are two basic types of two-dimensional model: vertically averaged and laterally averaged.

In vertically averaged models, concentrations are assumed to be uniform over the water column with variations oc-

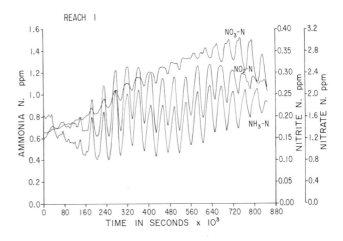

FIGURE 8.10 NH$_3$-N, NO$_2$-N, and NO$_3$-N variations two miles downstream of Blue Plains waste-treatment plant, July 15–24, 1969.

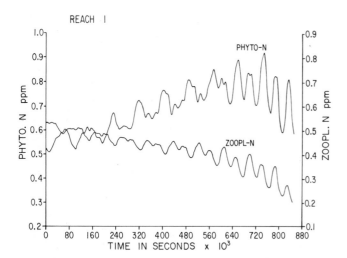

FIGURE 8.11 Phytoplankton-N and zooplankton-N variations two miles downstream of Blue Plains waste-treatment plant, July 15–24, 1969.

curring laterally in the plane of the water surface. Relatively few estuaries fall into this category; those that do tend to be wide and shallow and have circulation patterns influenced by Coriolis effects. Coastal embayments are usually treated by two-dimensional laterally averaged models. In these models, the transport processes are described by two-dimensional hydrodynamic models that include surface wind stress, bottom friction, and Coriolis accelerations. Examples of water-quality models in this category are given by Chen and Orlob (1971), Leendertse and Liu (1974), Brandes and Masch (1975), and Leimkuhler *et al.* (1975).

In laterally averaged models, concentrations are assumed to be uniform laterally, with variations occurring in the depth direction. Relatively narrow estuaries having a high degree of salinity stratification fall into this category, and most of the existing models are limited to descriptions of salinity and velocity variations with depth. Examples include Hamilton (1975) and Blumberg (1975).

The general state of development of two- and three-dimensional water-quality models is inherently limited by the availability of synoptic data for model verification.

CURRENT PROBLEMS AND RESEARCH NEEDS

It has been demonstrated that there is a high degree of coupling between intratidal cycle transport and mixing and biochemical transformation processes in estuaries. Gross simplification of the transport, as is the case with nontidal or through-flow water-quality models, leads to large differences in predicted nutrient concentrations even though identical biochemical components and rate constants are employed in the real-time and through-flow models.

Obviously much more is known about hydrodynamic

transport processes than is known about biochemical processes. Because of this, it is important that water-quality models incorporate correct transport processes so as to avoid obscuring further understanding of the complex biochemical transformations.

Large temporal variations of nutrient concentrations have been shown to occur within a tidal period at fixed locations. This suggests that a re-evaluation of traditional estuarine field data-collection techniques is necessary. More attention must be given to the determination of temporal as well as spatial variations of constituents in estuarine water-quality surveys. Relatively little useful information can be expected from the random collection of samples from a moving boat.

Probably the greatest limitation to the continued development and improvement of predictive water-quality models is the general lack of coordination between data needs for model verification and the data actually obtained in field surveys. Chemical and bioassays of water-quality samples should be made in relation to the structure of the water-quality model. In addition, basic laboratory experiments should be designed, again in relation to the model structure, to investigate biochemical transformation under controlled conditions.

ACKNOWLEDGMENTS

The major portion of this research was supported by the U.S. Environmental Protection Agency through Grant No. R800429 and was conducted under the supervision of the writer by Tavit O. Najarian. The grant was administered by the National Coastal Pollution Research Program, Corvalis, Oregon. The writer is grateful to Raymond A. Ferrara and Aldo Alvarez, Research Assistants in the R. M. Parsons Laboratory, for the simulation runs on the idealized estuary.

REFERENCES

Blumberg, A. F. (1975). A numerical investigation into the dynamics of estuarine circulation, Chesapeake Bay Inst. Tech. Rep. No. 91, Johns Hopkins U., Baltimore, Md.

Brandes, R. J., and F. D. Masch (1975). Estuarine ecologic simulations, *Symposium on Modeling Techniques,* Vol. I, Am. Soc. Civil Eng.

Chen, C. W., and G. T. Orlob (1971). Ecologic simulation for aquatic environments, OWRR Proj. IV.C-2044, Office of Water Resources Research, Dept. of the Interior, Washington, D.C.

Dailey, J. E., and D. R. F. Harleman (1972). Numerical model for the prediction of transient water quality in estuary networks, Tech. Rep. No. 158, Ralph M. Parsons Laboratory for Water Resources and Hydrodynamics, Dept. of Civil Engineering, MIT, Cambridge, Mass.

Dugdale, R. C., and J. J. Goering (1967). On uptake of new and regenerated forms of nitrogen in primary productivity, *Limnol. Oceanog. 12,* 196.

Dugdale, R. C., and J. J. MacIsaac (1971). On a computation model for the uptake of nitrate in the Peru upwelling region, *Invest. Pesquera 35,* 299.

Hamilton, P. (1975). A numerical model of the vertical circulation of tidal estuaries and its application to the Rotterdam waterway, *Geophys. J. R. Astron. Soc. 40*, 1.

Harleman, D. R. F., and C. H. Lee (1969). The computation of tides and currents in estuaries and canals, Tech. Bull. No. 16, Committee on Tidal Hydraulics, Corps of Engineers, U.S. Army, Vicksburg, Miss. See also D. R. F. Harleman and M. L. Thatcher, Appendix A: A User's Manual, May 1973.

Harleman, D. R. F., D. N. Brocard, and T. D. Najarian (1972). A predictive model for transient temperature distributions in unsteady flows, Tech. Rep. No. 175, Ralph M. Parsons Laboratory for Water Resources and Hydrodynamics, Dept. of Civil Eng., MIT, Cambridge, Mass.

Harrison, W. G., and J. E. Hobbie (1974). Nitrogen budget of a North Carolina Estuary, Rep. No. 86, Water Resources Research Inst. U. of North Carolina, Raleigh, N.C.

Holley, E. R., D. R. F. Harleman, and H. B. Fischer (1970). Dispersion in homogeneous estuary flow, *Am. Soc. Civil Eng. J. Hydraulics Div. 96*, HY8, 1691.

Jaworski, N. A., L. J. Clark, and K. D. Feigner (1971). A water resource-water supply study of the Potomac Estuary, Tech. Rep. No. 35, Environmental Protection Agency, Water Quality Office, Washington, D.C.

Leendertse, J. J., and S. K. Liu (1974). A water-quality simulation model for well-mixed estuaries and coastal seas: Vol. VI, Simulation, observation, and state estimation, New York City Rand Inst. Rep. R-1586-NYC, New York.

Leimkuhler, W., J. J. Connor, J. D. Wang, G. Christodoulou, and S. L. Sundgren (1975). Two-dimensional finite element dispersion model, *Symposium on Modeling Techniques*, Vol. II, Am. Soc. Civil Eng.

Najarian, T. O., and D. R. F. Harleman (1975). A nitrogen cycle water quality model for estuaries, Tech. Rep. No. 204, Ralph M. Parsons Laboratory for Water Resources and Hydrodynamics, Dept. of Civil Eng., MIT, Cambridge, Mass.

Quinlan, A. V. (1975). Design and analysis of mass conservative models of aquatic ecosystems, Ph.D. Thesis, Dept. of Civil Eng., MIT, Cambridge, Mass.

Thatcher, M. L., and D. R. F. Harleman (1972a). A mathematical model for the prediction of unsteady salinity intrusion in estuaries, Tech. Rep. No. 144, Ralph M. Parsons Laboratory for Water Resources and Hydrodynamics, Dept. of Civil Eng., MIT, Cambridge, Mass.

Thatcher, M. L., and D. R. F. Harleman (1972b). Prediction of unsteady salinity intrusion in estuaries: Mathematical model and user's manual, Tech. Rep. No. 159, Ralph M. Parsons Laboratory for Water Resources and Hydrodynamics, Dept. of Civil Eng., MIT, Cambridge, Mass.

Nutrient and Particulate Matter Budgets in Urban Estuaries

9

H. JAMES SIMPSON, SUSAN C. WILLIAMS,
CURTIS R. OLSEN, and DOUGLAS E. HAMMOND
Columbia University

INTRODUCTION

Estuaries have historically been common sites for growth of large population densities. Located at the interface between the sea and rivers draining interior continental regions, estuaries are ideal for development of commercial centers. In addition to providing natural harbors for ships, estuaries have frequently served as both food sources and sewage disposal systems for cities. Modern industrial civilization has placed additional demands on estuarine waters, including cooling for power plants, disposal of toxic industrial wastes, and recreational activities such as swimming and fishing.

One major problem of intensified use of estuarine waters is reduced dissolved oxygen levels. Sewage and other organic wastes discharged into estuaries increase the activity of the bacterial community as oxygen consumers. If the organic loading rate is sufficiently high to overwhelm natural oxygen resupply processes, dissolved oxygen can be completely consumed. The engineering and estuarine geophysics communities have made important contributions toward managing oxygen resources of urban estuarine waters. Mathematical models of estuarine oxygen distributions aid in the design and location of sewage treatment facilities for reducing oxygen demand to more acceptable levels and provide insight into natural capacities of various systems for coping with the oxygen demands they receive (O'Connor, 1970).

NUTRIENTS IN URBAN ESTUARIES

In recent years, substantial increases have occurred in algal populations in lakes, including large systems such as Lake Erie. These increases are believed to result primarily from the loading of additional plant nutrient elements—phosphorus and nitrogen. When excess algal growth can be directly related to man's activities, the changes are often ascribed to a loosely defined process called "cultural eutrophication." In most cases the most practical management tool for decreasing algal standing crops in lakes is to reduce the input of phosphate. As a result, the general policy direction for nutrient management in several European countries and North America is to decrease the phosphate levels in detergents, to construct tertiary treatment facilities for phosphate removal from sewage whenever possible, and/or to divert sewage discharge from lakes if an acceptable alternative is available.

Excess algal growth problems have also developed in some large estuarine systems, including the Potomac downstream of Washington, D.C., and removal of both phosphorus and nitrogen from sewage effluent has been proposed in several cases. Because of the costs of tertiary sewage-treatment facilities and problems of sludge disposal from such facilities, it is important to examine critically the benefits of nutrient removal for estuaries on a case-by-case basis.

Mathematical representations of algal populations, nutrients, and even higher trophic levels such as zooplankton and fish can be constructed in a similar way to those generated to describe the balance of dissolved oxygen in estuaries. The predictive capability of such models is sometimes not clear, since the response of phytoplankton communities, as well as higher trophic levels, to environmental parameters is more complex than for bacteria in their role as oxygen consumers. In light of large uncertainties in critical biological parameters, management alternatives can sometimes be reasonably examined using extremely simplified conceptual models of water circulation and nutrient–algae interactions. The approach described here (see also Simpson *et al.*, 1975) is to examine the distribution of phosphate in the Hudson Estuary and the rate of loading from sewage outfalls, in terms of a very simplified description of water circulation and phosphate behavior in the harbor region adjacent to New York City. Phosphate is the nutrient most frequently considered for nutrient removal from sewage effluent and is somewhat simpler to treat in estuarine budgets than nitrogen. Nutrient distributions in the Hudson are compared with those in San Francisco Bay to indicate the first-order similarities of nutrient–algae relationships in these two systems.

SEWAGE AND PHOSPHATE IN THE HUDSON ESTUARY

The Hudson Estuary, portions of which are entirely surrounded by the New York City metropolitan area, is an example of an estuarine system heavily loaded with sewage and other discharges that substantially alter the ambient water quality. Dissolved oxygen levels of less than 40 percent of saturation with atmospheric oxygen concentrations are not uncommon during summer months. Sewage is discharged to the Hudson Estuary near New York City from about a dozen major treatment-plant outfalls and numerous smaller plants with a combined flow of approximately 80 m^3/sec ($\sim$1.8 billion gallons/day) plus a number of raw sewage outfalls, largely from Manhattan Island, which total about 20 m^3/sec ($\sim$0.5 billion gallons/day). Approximate locations of the largest outfalls are shown in Figure 9.1. Of the major treatment plant outfalls, about two thirds of the total volume is discharged from secondary treatment plants operated by New York City, and the remainder comes from primary treatment plants mostly from New Jersey.

The major secondary treatment plants in New York City,

ranging in age from a few years to about 40 years, are reasonably efficient at reducing biological oxygen demand of sewage. Except for storm runoff periods, when treatment plants are essentially bypassed, about two thirds of the oxygen demand of the sewage is removed. Construction of major facilities is now under way to treat most of the remaining raw discharges, which currently contribute on the order of one third of the total dry weather biological oxygen demand to the Lower Hudson Estuary. Primary treatment outfalls, mostly from New Jersey, discharge about one third of the total treated sewage flow and about one half of the total oxygen demand. Thus in terms of the current oxygen budget, the largest impact of sewage discharge is from partially treated New Jersey outfalls. The oxygen demand from these sources will dominate to an even greater extent when the current construction projects in New York City provide secondary treatment for the raw discharge from Manhattan.

All the major discharges of sewage to the Lower Hudson Estuary supply large amounts of primary plant nutrients, especially ammonia and phosphate, as well as organic materials, which constitute the major load on the dissolved oxygen resources of the water. The supply rate of nutrients is more directly proportional to volume of sewage flow than is biological oxygen demand, since present treatment operations (raw, primary, or secondary) do not have nearly so great an effect on nutrients as they do on dissolved and suspended organic carbon. It is important to establish to what extent these sewage-derived nutrients are converted to organic matter by phytoplankton and how this affects oxygen concentrations and other water-quality factors.

The concentrations of dissolved phosphate as well as other primary nutrients in the harbor region of the Hudson Estuary are approximately two orders of magnitude higher than those usually considered as limiting to phytoplankton growth. Phosphate concentrations do not vary greatly from one location to another in the harbor, despite the presence of a number of large discrete sources. The smoothness of observed phosphate concentrations indicates the effectiveness of mixing by tidal currents and density-induced nontidal estuarine circulation. In general, the mean values of phosphate in the Inner Harbor range between 2×10^{-6} grams-atoms/liter during spring high freshwater runoff and 6×10^{-6} gram-atoms/liter during summer low freshwater discharge. To illustrate the general character of observed data, phosphate and salinity measurements from water samples collected along the axis of the Hudson Estuary are shown for several representative time periods (Figure 9.2).

Total phosphorous measurements, as well as measurements of particulate organic phosphorus and dissolved organic phosphorus concentrations, were made for most of the transects shown in Figure 9.2. In general, about one half to two thirds of the total phosphorus in the water was present as molybdate-reactive phosphate, with most of the remainder as particulate organic phosphorus. During high freshwater flow (March and April) the fraction of particulate organic phosphorus was somewhat greater. The data shown are for samples

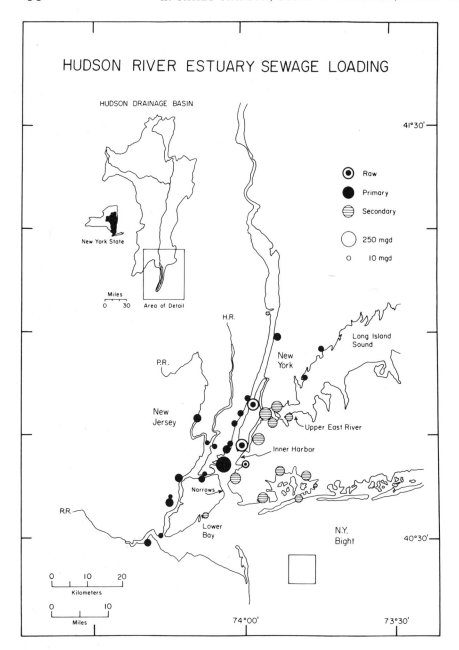

FIGURE 9.1 Location map for major sewage outfalls to the Hudson River Estuary: the total discharges shown are about 2.3 billion gallons per day ($\sim$100 m³/sec). The areas of the circles are proportional to the volume of sewage outfall at that discharge site. The only outfalls included are those in excess of 10 million gallons per day. The enclosed square in the New York Bight indicates the approximate location of sewage sludge and dredge spoil dumping for the New York area. Abbreviations R.R., P.R., and H.R. indicate small tributary rivers (Raritan, Passaic, and Hackensack Rivers).

collected along the axis of the Hudson, through the middle of the Inner Harbor, the Narrows, the Lower Bay, and the apex of the New York Bight (Figure 9.1). Plots of data from samples collected in heavily loaded and more restricted zones of the Inner Harbor such as the East River are greater than the trends shown, sometimes by as much as a factor of 2 in phosphate concentration, but the data shown are typical of most of the harbor volume.

Upstream of New York Harbor, phosphate and salinity decrease at approximately the same rate until freshwater is reached, where the phosphate levels remain nearly constant at 0.5 to 1.0 μm/liter, depending on the freshwater flow rate. The covariance of salinity and phosphate suggests that removal processes for phosphate such as algal uptake do not dominate phosphate concentrations in the salt intruded reach of the Hudson and that physical transport of the water by estuarine circulation is most critical in establishing the phosphate distribution.

The geometry of sewage loading is presented schematically in Figure 9.3. Most of the sewage is discharged directly to the Inner Harbor, with less than 5 percent of the total added upstream of the Inner Harbor (segments A, B, C, and D of

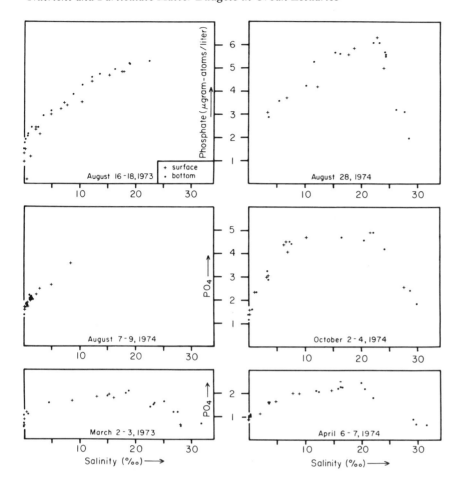

FIGURE 9.2 Phosphate concentrations as a function of salinity in the Hudson River Estuary: freshwater flows vary from <300 m³/sec for the August profiles to ⌐450 m³/sec for October and to ⌐1250 m³/sec for March and April. Phosphate is higher during lower flow rates. Phosphate data were not obtained at higher salinities during two of the August surveys.

Figure 9.3). Saline water and sewage phosphate added to the Inner Harbor are spread well upstream of the discharge sites by estuarine circulation and as far upstream as segment D during low flow periods.

To illustrate the relative magnitudes of phosphate source terms to the Inner Harbor and provide some sense of the time scale of removal by estuarine circulation, a one-box model for phosphate in the Inner Harbor can be constructed (Figure 9.4). The mean concentration of phosphate in the Inner Harbor, multiplied by the volume of this portion of the estuary, gives the total amount of phosphate in solution. Inputs of phosphate are (1) direct sewage discharge, (2) diffusion from the sediments, (3) downstream supply from the freshwater reach of the Hudson, and (4) particulate oxidation in the water column, with direct sewage discharge supplying about three fourths of the total. Dividing the total quantity of dissolved phosphate in the Inner Harbor by the supply rate indicates a residence time for phosphate of 2 days during high flow and 7 days during low flow.

A similar calculation can be done using a first-order estimate of the outflow rate of phosphate from the Inner Harbor. Removal of phosphate is predominantly through the Narrows, into the Lower Bay, and then to the adjacent coastal waters of the New York Bight. Some phosphate discharge does occur through the Upper East River into western Long

Island Sound. Based on estimates of the magnitude of this transport, we have decreased the loading rate of phosphate to the Inner Harbor used in Figure 9.4 by an equivalent amount—roughly 20–25 percent of the discharge rate through the Narrows. Water flowing out through the Narrows in the upper half of the water column has a mean phosphate concentration similar to that in the Inner Harbor, while that flowing upstream in the lower layer has a lower phosphate concentration and a higher salinity. Using a two-layer advective transport calculation, based on field salinity and phosphate concentrations during both high and low flow conditions, the removal rate of phosphate is comparable with that estimated in Figure 9.4 for the input rate. Thus to the first approximation, sewage phosphate is discharged to the lower Hudson into waters with salinities averaging about two thirds of seawater salinity, mixed reasonably well through the Inner Harbor over a time period averaging between 4 and 14 tidal cycles, and discharged from the system by estuarine circulation.

If this description is reasonable, then there are substantial implications for phosphate management policies in the New York City area. The primary purpose of decreasing phosphate discharge to receiving waters is to reduce adverse effects resulting from excess algal growth in the receiving water. The Hudson Estuary upstream of the Narrows in

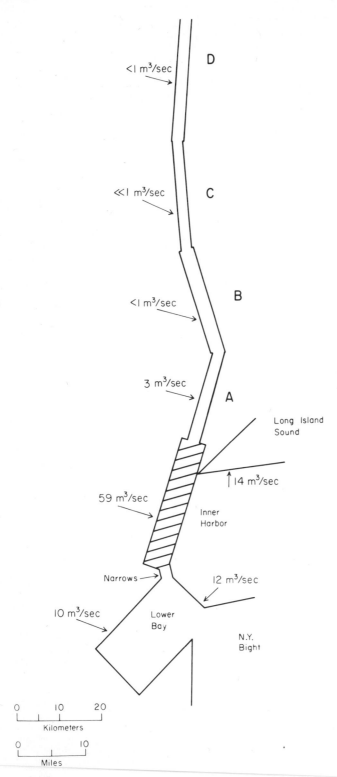

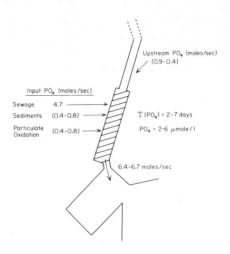

FIGURE 9.4 Approximate fluxes and concentrations of phosphate in the Inner Harbor: sewage discharge is assumed constant seasonally, while the other fluxes vary somewhat from high-flow ($\sim$1200 m³/sec), low-temperature ($\leqslant$5°C) conditions listed first in parentheses to low-flow ($\leqslant$300 m³/sec), high-temperature ($\leqslant$20°C) conditions. The major cause of changes in Inner Harbor phosphate residence times (Υ) is change in rate of freshwater discharge.

the Inner Harbor has a large surplus of nutrients for algal growth. The algal population within this reach of the Hudson is surprisingly low. The maximum warm-season standing crops of algae are on the order of only 10 percent of those of Lake Erie, a system that is generally accepted as having substantial areas that are clearly eutrophic. During winter months, algal standing crops in the Hudson are lower than in summer months by more than an order of magnitude. Thus the Hudson Estuary is not now characterized by nuisance growth of algae. In addition, algal activity is not currently limited by available nutrients. Large excesses of ammonia and nitrate as well as phosphate are present. Thus reduction of the current discharge rate of phosphate by tertiary treatment of sewage for phosphate removal probably would have little effect on the activity of algae in this reach of the Hudson.

A first-order budget for sewage-derived nitrogen in the Hudson has been presented (Garside et al., 1976), which also indicates that dissolved nitrate and ammonia are removed from the Hudson Estuary predominantly by estuarine circulation and not by phytoplankton uptake. There seems little indication of nutrient limitation being significant for the Inner Harbor area of the Hudson Estuary.

The factors that currently limit phytoplankton standing crops in the Hudson are not well defined, but growth rates have been clearly shown to be light- and temperature-regulated (Malone, 1977). The most reasonable factors determining standing crops in the Hudson are light limitation due to the high suspended solids content of the water and rapid removal of algal cells from the estuary upstream of the Narrows

FIGURE 9.3 Schematic loading of sewage to segments of the Hudson: the mean annual flow of the Hudson is $\sim$550 m³/sec, with saline water intruding to region A during high flow and to region D during low flow. The total mean annual freshwater flow from other tributaries to the harbor complex, the Passaic, Hackensack, and Raritan Rivers (see Figure 9.1), is about 80 m³/sec.

by estuarine circulation. Neither of these factors could be readily affected by changes in sewage-treatment processes since most of the suspended particulates are not supplied directly from sewage outfalls. If a substantial reduction in suspended solid levels during low-flow summer months were to occur, then algal growth rates would increase, perhaps reaching nuisance levels. This possibility should be carefully examined but does not seem of primary concern for the next decade.

After leaving the Inner Harbor through the Narrows, the high-nutrient estuarine waters support increasing phytoplankton standing crops, especially after reaching the adjacent coastal waters of the New York Bight. The magnitude of this increased phytoplankton activity can be clearly seen in higher chlorophyll contents in the New York Bight and even in the Lower Bay. Growth rates of phytoplankton in the nearshore waters are primarily light- and temperature-regulated and not nutrient-limited (Malone, 1976) and thus not very sensitive to control by nutrient discharge control upstream within the estuarine region.

The total phytoplankton biomass of the coastal waters adjacent to the Hudson is, however, related to the quantity of nutrients reaching the area. Coastal waters adjacent to large rivers always support enhanced phytoplankton activity produced by injection of primary nutrients and usually support greater fish populations as a result. Dissolved oxygen in bottom waters of small areas of the apex of the New York Bight approach values comparable with those of the Inner Harbor (<40 percent of saturation). Bottom waters of the New York Bight well beyond the zone of influence of the Hudson Estuary also show substantial oxygen depletion during the summer. The critical parameters for control of dissolved oxygen of bottom waters immediately offshore of the Hudson are still not well defined at this time. It is thus not clear at present if any major detrimental effects to the coastal waters are produced by enhanced nutrient discharge from the Hudson Estuary.

In light of the current circumstances of sewage-treatment, nutrient-content, and water-quality parameters in the Hudson Estuary, the most reasonable policy for management of nutrient discharge during the next decade would not appear to be the initiation of tertiary treatment for phosphate removal (or for other primary nutrients). Considering the low summer dissolved oxygen levels within the Inner Harbor, the major efforts in sewage treatment should be to complete the secondary treatment facilities under construction in New York City and to upgrade the New Jersey discharges substantially to include secondary treatment. Such an effort would require a minimum of a decade, considering the current mix of political interests, but should significantly improve the dissolved oxygen content of the most heavily impacted region of the Hudson Estuary. During that period, the question of nutrient removal and its potential effect on coastal water quality could be more extensively examined from the viewpoint of possible construction of tertiary sewage treatment facilities.

The conclusions presented here are based on extremely simple conceptual models of estuarine circulation, which do not include any details of circulation or flow dynamics. They involve some first-order approximations of net transport rates and time scales of flushing, as well as field data on the distribution of phosphate and chlorophyll as a function of salinity and location within the Hudson. Numerical models, more intensive programs of field measurements, or both can certainly provide additional insights into the behavior of nutrients within the Hudson, but it is unlikely *from a management viewpoint* that major changes in the conclusions about phosphate behavior within the Inner Harbor of the Hudson Estuary would result from more elaborate model descriptions of the circulation or from a more detailed knowledge of the distribution of phosphate from field data.

SILICATE IN URBAN ESTUARIES

In open-ocean surface waters, dissolved silicate is often reduced to extremely low concentrations by diatoms, a significant group of primary producers in the phytoplankton community. In rivers and estuaries, there is usually sufficient silicate from weathering of silicate rocks to maintain a substantial excess over the requirements of diatoms. Many estuarine studies have concentrated on establishing whether dissolved silicate concentrations within the zone of mixing between freshwater and a saline end member deviate significantly from a conservative mixing line (Boyle *et al.*, 1974). In cases where significant negative deviations occur, the removal process has usually been attributed to diatom uptake (Wollast and DeBroeu, 1971) or inorganic silicate reactions involving clay minerals in the suspended particulates (Bien *et al.*, 1958). In the case of both the Hudson Estuary (Simpson *et al.*, 1975) and the San Francisco Bay Estuary (Peterson *et al.*, 1975), there seems clear evidence that biological removal is the dominant process extracting silicate from these estuarine waters. In the case of the Hudson, there is evidence of a significant extra source of silicate within the estuary from the discharge of sewage, primarily because of the large quantity of sewage discharged and the relatively low ambient silicate concentrations in the most heavily loaded reach of the estuary. In San Francisco Bay, the quantity of sewage discharge is lower, and the river end-member silicate concentration is much higher (see Figure 9.5).

In estuaries where diatoms are able to affect dissolved silicate concentrations significantly, as in the case of the Hudson River and San Francisco Bay, the distribution of silicate could potentially provide valuable insights into rates of activity and distribution of phytoplankton communities. Nutrients such as phosphate and nitrate appear to have reasonably fast recycling times in estuaries, whereas silicate appears to be regenerated at much slower rates. Thus, dissolved silicate concentrations could provide some indication of integrated phytoplankton activity as water passes through an estuarine system. The relative importance of diatoms and phytoplankton that are not major

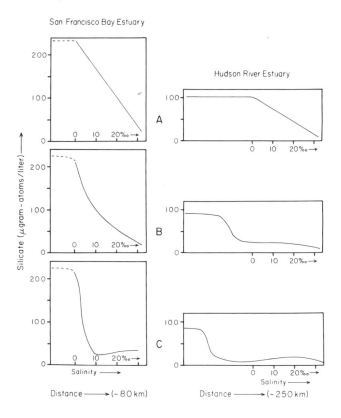

FIGURE 9.5 Silicate variations along the axes of two estuaries: transects are schematic, adapted from data in Peterson *et al.* (1975) and Simpson *et al.* (1975). Conditions vary from high-flow, low photosynthetic activity profiles (A) to intermediate (B) and low-flow, high photosynthetic activity profiles (C). The Hudson has an extremely long tidal freshwater reach in which diatoms remove appreciable silicate from solution during low-flow, warm months.

consumers of silicate must be established by field measurements before dissolved silicate trends could become useful as a quantitative indicator of total phytoplankton activity. In San Francisco Bay, diatoms are the dominant phytoplankton group (Peterson *et al.*, 1975), whereas in the Hudson there is a smaller proportion of diatoms in the Inner Harbor during summer months of highest productivity (Malone, 1977). Slow regeneration of silicate within the water column and release of silicate from sediment pore waters would reduce silicate depletion rates; but these effects can possibly be estimated more accurately than can more poorly defined, but crucial, regeneration rates of phosphate, nitrate, and ammonia. Thus, silicate appears to be promising as a chemical tracer in estuarine systems, which could be used to provide better definition of critical rates in nutrient and phytoplankton estuarine models.

SEDIMENTATION IN ESTUARINE HARBORS

Another problem of management of estuaries in urban regions is that of suspended particulates. High suspended

loads can degrade both recreational values and food resources of estuarine waters and affect navigation uses in rapid shoaling areas. Estuaries are often sites of high sediment deposition, which conventional wisdom usually ascribes to flocculation of suspended particulates in the presence of saline water, combined with variations in current velocity associated with alternating tidal currents and a two-layer estuarine circulation pattern. The net upstream current direction in the bottom layer tends to prevent the removal of suspended particulates that settle down from the surface layer. In estuarine harbors, the natural tendency for sedimentation can be enhanced or otherwise altered by deepening of the channel, construction of piers and barriers to wave action, and diversion of freshwater into or from the system. Substantial changes in accumulation patterns and rates of sedimentation have occurred over periods of a few decades because of man's activities in a number of estuaries, including the Mississippi River mouth, the lower Savannah River, and Charleston Harbor (Simmons and Herrmann, 1972). Dredging of large amounts of rapidly accumulating sediments is often necessary to maintain navigation depths.

The Hudson Estuary is no exception in terms of shoaling problems. Approximately 4 million metric tons (dry weight) of dredged sediment are removed annually from New York Harbor and discharged at a site offshore in the New York Bight (Gross, 1972). A number of proposals for alteration of harbor geometry by dams, extensive dredging operations, or both have been suggested by the Corps of Engineers on the basis of physical model studies (Panuzio, 1965). The purpose of the alterations would be to reduce the current shoaling rate and thus eliminate some dredging.

Over the past several years, we have been mapping the distribution of recent sediment accumulation within the Hudson Estuary (see Figure 9.6). The technique we have exploited is to measure in sediments the radionuclides derived from global fallout as well as from low-level releases of a nuclear power reactor located near the upstream end of the saline intrusion in the Hudson. Cesium-137, cesium-134, and cobalt-60 are present in Hudson sediments at activity levels comparable with natural radioactivity and provide a sensitive indicator of sediments accumulated within the last 10–15 years.

On the basis of our mapping, sediment accumulation rates of much less than a few centimeters per year are typical of much of the total area Hudson Estuary. There are restricted regions of rapid accumulation that appear to account for most of the total deposition. Shallow marginal coves, mostly located well upstream of the Inner Harbor, have sedimentation rates of several centimeters per year. The dominant zone of deposition, however, is within the harbor, where large areas have accumulation rates on the order of 5–10 cm/yr, and at least one area has shoaling rates of 10–30 cm/yr. Sediments accumulating in the harbor have radionuclides present that could have only come in significant amounts from the reactor located approximately 60 km upstream. Thus fine-grained sediments now accumulating in the harbor must have a major component

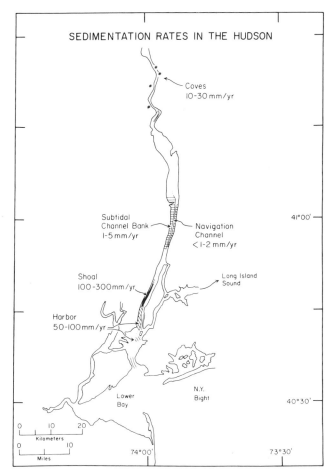

SEDIMENTATION RATES IN THE HUDSON

FIGURE 9.6 Sedimentation rates in the Hudson: central channel areas in the Narrows, along Manhattan, and upstream are frequently scoured by tidal currents and have little sediment accumulation while upstream protected coves and large areas of the main harbor have high accumulation rates. The reactor site from which radioactive releases have tagged sediments now found in the harbor is located about 15 km downstream of the cove site near the top of the map.

of particulates that came from well upstream of the harbor area.

During high freshwater runoff periods, when the suspended load of the Hudson is substantially higher than during low flow, saline water in the Hudson Estuary is displaced downstream to within 15 km or less of the Inner Harbor. Thus the first contact of suspended particulates with saline water occurs reasonably close to the region where high accumulation rates are observed. In addition, the cross-sectional area of the Hudson Estuary increases abruptly by a factor of about 2 at the southern tip of Manhattan, which decreases the net downstream advection rate about midway through the Inner Harbor. Another physical anomaly occurs near the upstream end of Manhattan, where large quantities of construction rubble were dumped in a deep narrow channel between 1940–1945,

reducing the total cross-sectional area by about a factor of 2 over a short reach of the estuary. This is believed to enhance vertical exchange greatly and to destroy partially the two-layer circulation in that area. This change in channel area has been invoked to explain why dredge spoil removal from the navigation channels of the Inner Harbor between 1945 and 1960 was about twice the rate for the 15 years preceding 1940 (Panuzio, 1965). Another factor is the presence of a large number of piers in the Inner Harbor, which provide local sites of very high accumulation as tidal currents are reduced to a small fraction of the main-channel velocities.

There appear to be several factors associated with the location of New York Harbor that are favorable for rapid accumulation of sediment. There do not seem to be any simple or reasonable physical changes within the Inner Harbor that could be made to reduce the shoaling problem greatly. One possible direction that could aid the control of shoaling would be to reduce the supply of suspended particulates to the Hudson River. Currently, the sources of particulates are poorly defined but appear to be located well upstream of the Inner Harbor. It is highly probable that a major reduction of the suspended load within the Hudson Estuary would enhance phytoplankton activity and thus possibly degrade some aspects of water quality within the Hudson Estuary below the present levels.

Sediment accumulation in the San Francisco Bay Estuary based on budget calculations (Conomos and Peterson, 1974, 1976) is about the same as the rate of dredge spoil removal from the Hudson (4×10^6 tons/yr). The concentration of suspended material in the rivers discharging to San Francisco Bay is higher than for the Hudson, with a mean annual combined freshwater discharge of about 1.5 times that for the Hudson. The primary source of shoaling materials in San Francisco Bay is from the Sacramento and San Joaquin Rivers, which supply most of their particulates during high-flow winter months. During low-flow summer months, particulates that were brought in during higher flow periods are redistributed, especially upstream of the original deposition zone, as the intrusion of saline water progresses inland. The broad shallow morphology of San Francisco Bay favors remobilization of sediments both by strong tidal currents and by wind-induced turbulence (Conomos and Peterson, 1976).

FINAL COMMENTS

San Francisco Bay has a large and relatively isolated zone south of the main axis of estuarine circulation. Because of the relatively slow renewal rate of water in this large area, phosphate concentrations build up to values more than five times those in the Hudson during some seasons, despite much lower absolute loading rates. The general pattern of seasonal phosphate concentrations in South San Francisco Bay is inversely correlated to water renewal time based on salinity calculations, as would be expected if water circulation rates were a major removal process for dissolved phosphate. Flushing rate calculations using phosphate concentrations as a conservative tracer during one December–January period indicated about half of the water of South San Francisco Bay was renewed in one month, which was the same renewal rate computed from salinities (McCulloch et al., 1970).

Profiles of nutrients and salinity in San Francisco Bay indicate both positive and negative deviations of phosphate from a conservative mixing line (Conomos and Peterson, 1975), but the effects were smaller than for other nutrients. Significant depletion of nutrients is observed in zones of highest productivity near the turbidity maximum. San Francisco Bay currently does not have excessive phytoplankton concentrations nor does nutrient limitation of growth appear to be a significant factor in the general features of the phytoplankton ecology of the bay·(Peterson et al., 1975; Peterson and Conomos, 1975).

There are strong similarities in some aspects of the behavior of phosphate and silicate in the Hudson Estuary and the San Francisco Bay Estuary and in the magnitudes of suspended sediment accumulation. The details of geometry of sediment accumulation and sewage loading are obviously greatly different, but there are enough parallels to suggest some insights into modeling and management problems in other heavily affected urban estuaries.

In the case of the Hudson, based on the simple conceptual models discussed here, management of both nutrients and sediments in the New York City area are very strongly affected by the physical circumstances of the region of mixing of freshwater and saline water. Estuarine circulation appears to have somewhat simplified the task of managing sewage treatment because of the rapid transport and removal of nutrients from the Inner Harbor. In the case of sediment accumulation, however, the pattern of estuarine circulation in the Hudson has helped to produce a zone of rapid shoaling that coincides with the area that is most intensively used by man. In this situation, there do not seem to be any reasonable near-term management policies that could significantly reduce the magnitude of dredging required.

RECOMMENDATIONS FOR FUTURE WORK

Many management decisions for urban estuaries must be made on the basis of inadequate information. Ideally one would like to increase the state of knowledge of estuarine systems to the point where detailed predictive numerical models were available to describe the behavior of nutrients, sediments, and many other water-quality parameters in time steps much shorter than a tidal cycle. That is a distant goal, and our predictive ability may ultimately be limited by the complexity and transient-dominated nature of estuarine biological, chemical, and physical processes. Currently, there is much less information about biology, chemistry, and particulate phases in estuaries than about physical transport of water. Thus, one of the crucial directions for future work is greater integration of estuarine physical transport modeling and field programs with other areas of estuarine research. In many cases, it should be possible to make better management decisions if insights from several fields are pooled in relatively simple models of estuarine processes.

ACKNOWLEDGMENTS

Financial support for research on the Hudson Estuary reported here was provided by the Environmental Protection Agency (R803113-01, 02) and the Energy Research and Development Administration (E(11-1)-2529).

REFERENCES

Bien, G. S., D. E. Contois, and W. H. Thomas (1958). The removal of soluble silica from fresh water entering the sea, Geochim. Cosmochim. Acta 4, 35.

Boyle, E., R. Collier, A. T. Dengler, J. M. Edmond, A. G. Ng, and R. F. Stallard (1974). On the chemical mass balance in estuaries, Geochim. Cosmochim. Acta 38, 1719.

Conomos, T. J., and D. H. Peterson (1974). Biological and chemical aspects of the San Francisco Bay turbidity maximum, Mem. Inst. Geol. du Bassin d'Aquitaire 7, 45.

Conomos, T. J., and D. H. Peterson (1975). Longitudinal distribution of selected micronutrients in northern San Francisco Bay during 1972, in Proceedings of a Workshop on Algae–Nutrient Relationships in the San Francisco Bay and Delta, The San Francisco Bay and Estuarine Assoc., pp. 127–146.

Conomos, T. J., and D. H. Peterson (1976). Suspended-particle transport and circulation in San Francisco Bay: An overview, Proceedings of the 3rd International Estuarine Research Federation Conference, in Estuarine Processes, 2, 82–97.

Garside, C., T. C. Malone, O. A. Roels, and B. C. Sharfstein (1976). An evaluation of sewage-derived nutrients and their influence on the Hudson Estuary and New York Bight, Estuarine Coastal Marine Sci. 4, 281.

Gross, M. G. (1972). Geologic aspects of waste solids and marine waste deposits, New York metropolitan region, Bull. Geol. Soc. Am. 83, 3163.

Malone, T. C. (1976). Phytoplankton productivity in the apex of the New York Bight: Environmental regulation of productivity/chlorophyll a, in The Middle Atlantic Continental Shelf and New York Bight, Limnology and Oceanography Special Symposium, Vol. 2, M. G. Gross, ed., Allen Press, Lawrence, Kansas, pp. 260–272.

Malone, T. C. (1977). Environmental regulation of phytoplankton productivity in the Lower Hudson Estuary. Estuarine Coastal Marine Sci. 5, 157.

McCulloch, D. S., D. H. Peterson, P. R. Carlson, and T. J. Conomos, (1970). A preliminary study of the effects of water circulation in the San Francisco Bay Estuary, U.S. Geological Survey Circ. 637A, Washington, D.C.

O'Connor, D. J. (1970). Water quality analysis for the New York Harbor complex, in Water Pollution in the Greater New York Area, Gordon and Breach, New York, pp. 121–144.

Panuzio, F. L. (1965). Lower Hudson River siltation, in Proceedings of the Federal Interagency Sedimentation Conference, Misc. Pub. 970. U.S. Dept. of Agriculture, Washington, D.C., pp. 512–550.

Peterson, D. H., and T. J. Conomos (1975). Implications of seasonal chemical and physical factors on the production of phytoplankton in northern San Francisco Bay, in *Proceedings of a Workshop on Algae–Nutrient Relationships in the San Francisco Bay and Delta,* The San Francisco Bay and Estuarine Assoc., pp. 147–166.

Peterson, D. H., T. J. Conomos, W. W. Broenkow, and E. P. Scrivani (1975). Processes controlling the dissolved silica distribution in San Francisco Bay, in *Estuarine Research,* Vol. I, Academic Press, New York, pp. 153–187.

Simmons, H. B., and F. A. Herrmann, Jr. (1972). Effects of man-made works on the hydraulic, salinity and shoaling regimens of estuaries, *Geol. Soc. Am. Memoir 133,* 555.

Simpson, H. J., D. H. Hammond, B. L. Deck, and S. C. Williams (1975). *Nutrient Budgets in the Hudson River Estuary,* Symp. Series No. 18, American Chemical Soc., Washington, D.C., pp. 618–642.

Wollast, R., and F. DeBroeu (1971). Study of the behavior of dissolved silica in the estuary of the Scheldt, *Geochim. Cosmochim. Acta 35,* 613.

Suspended Sediment Transport and the Turbidity Maximum

10

RONALD J. GIBBS
University of Delaware

INTRODUCTION

The phenomena of rivers discharging into estuaries are of interest to a wide variety of scientists and engineers for many reasons. The suspended materials that are discharged by rivers into estuaries and oceans transport many pollutants and are the natural material that fills our channels and harbors. From the biological and health viewpoint, the suspended materials are seen as the natural food of the filter feeding organisms; therefore, pollutants transported by the suspended material that is discharged can adversely affect many varieties of seafood. This is one reason why a knowledge of the phenomena of a river discharging into the ocean or an estuary is critical. If we understand the mechanisms of transport of the suspended material, we can intelligently estimate the fate of many pollutants. Possibly we can learn how to divert or decrease the detrimental turbidity in regions of seafood harvesting. Another advantage to be gained from knowledge in this area is the ability to maintain our navigable waterways at lower cost by utilizing dredging techniques compatible with natural processes.

PREVIOUS RESEARCH

In reviewing the progress made in studies of the transportation of suspended material in estuaries, we find that numerous researchers have worked on one aspect or another of suspended material. Several volumes provide a current view of the present status in this area and contain a wealth of references to work done on various aspects of the subject: Nelson (1973), Gibbs (1974), and Cronin (1975). In these volumes, the vast majority of studies on suspended material have dealt with some particular aspect of the problem, for example, trace-metal behavior, organic material, the fate of clay minerals. Most of these studies give little additional information usable in an understanding of other aspects of suspended material in estuaries. The number of suspended material investigations that have obtained information on the movement of the material in water is limited. Meade (1972) reviewed a number of suspended-material studies and discussed flocculation, the turbidity maximum, and the transport processes. Detailed studies on the turbidity maximum and on suspended material transport of the Chesapeake Bay and some of its tributaries are reported in a series of papers by Nichols and co-workers (1972, 1973) and by Schubel (1968, 1972, and 1974), making this probably the most extensively studied area along the eastern United States coastline. San Francisco Bay is the estuary on the west coast in which suspended material transport is best understood through the work of the U.S. Geological Survey, mainly by Conomos and Peterson (1974) and Peterson *et al.* (1975) along with the earlier work of Krone (1962, 1966). In Great Britain and in Europe, a number of estuaries have been intensively studied, providing an understanding of their suspended material transport.

SUSPENDED-MATERIAL INPUT

The input of suspended material to estuaries around the world is variable in concentration and composition from river to river, as well as changing with time in any particular river. The natural variability in the average concentration of suspended material delivered to estuaries can change from river to river by many orders of magnitude (Table 10.1). This variation in average concentration is related to seasonal changes in precipitation and runoff within the drainage basins of the rivers. The regions of extreme seasonal variations in precipitation and runoff also have the widest variation in suspended-material concentration, and the regions having reasonably steady climate have the least variation in suspended-material concentration. This relationship is illustrated in the Yukon River, with its extreme climatic range. The Yukon flow has a concentration of suspended material of over 1000 mg/liter at the time of spring ice break-up and a concentration of less than 1 mg/liter in the flow under the ice during the winter. On the other hand, the range is correspondingly reduced in the less variable climates even though they have high precipitation, as illustrated by some of the tributaries of the Amazon River, e.g., the Javari River, having a concentration of 81

ppm at the time of high discharge and a concentration of 40 ppm at the time of low discharge, and the Jutai River, having concentrations of 45 ppm and 33 ppm at times of high and low discharge, respectively (Gibbs, 1967). In the regions of the world having temperate climate, the seasonal concentrational variation is intermediate between these extremes, as illustrated by the Susquehanna River, the major source river for the Chesapeake Bay, having a normal seasonal variation of 10 to 140 ppm (Schubel, 1974). In most temperate regions of the world, episodic climatic events may have a tremendous effect on suspended material concentration. For example, following tropical storm Agnes, the concentration of suspended sediment in the Susquehanna River exceeded 10,000 mg/liter in its discharge entering the Chesapeake Bay (Schubel, 1974).

Considering solely the concentration of suspended material is not sufficient; the quantity of suspended material delivered to estuaries must also be considered. In Table 10.1 the suspended material discharge is given for each of those rivers for which suspended-material concentrations are given. The wide variation of these values is related to the volume of water discharged. As concentration of suspended material varies with seasonal climatic variation, so does the suspended-material discharge, with even greater variations. In the James River, a temperate climate river entering the Chesapeake Bay, this seasonal variation results in 90 percent of the yearly suspended material being discharged during 11 percent of the year (Nichols and Thompson, 1973). A more striking example of the nonuniformity of the discharge of suspended material is the case given by Schubel (1974), in which the Susquehanna River discharged more suspended sediments into the Upper Chesapeake Bay in a one-week period following storm Agnes than had been discharged during the past three decades and probably during the past half-century. Such seasonal and episodic variations illustrate the difficulties that can be encountered in attempting to sample suspended material. An indication of such episodic events can be determined by looking at the records of flooding, which have been recorded more extensively than have the suspended-material transport data. It appears that in most temperate regions these episodic events transport the bulk of the material over a long time span.

The effect of man on the increase of suspended-material concentration and discharge of rivers has been striking. The cutting of forests, plowing for farmland, and the multiplicity of construction projects associated with man's industrial and residential expansion have markedly increased the concentration and discharge of suspended material. In other regions, the effect is the opposite, as man has erected dams that trap sediments, thereby cutting off the supply. However, since transport/erosion processes continue unabated in these cases, a new equilibrium will be established usually after a great deal of erosion of the shoreline and bottom.

The composition of the suspended material discharged by rivers into the oceans varies from river to river, depending on the composition of the rocks and soils in the river's

TABLE 10.1 Concentration and Discharge of Sediment by River

River	Average Concentration Suspended Material, ppm[a]	Sediment Discharge, 10^{12} g/yr[b]
Mississippi	250	312
Susquehanna[c]	10–140	0.75
Nile	595	110
Rio Grande	42,600	8.5
Danube	350	19.4
Rhone	560	
Amazon	90	499
Congo	23	64.7
La Plata	430	

[a] From Garrels and Mackenzie (1971).
[b] From Holeman (1968).
[c] From Schubel (1974).

drainage basin, the weathering climate to which these rocks and soils have been exposed, and the energy of rivers to transport various sizes of material. The various minerals that make up the suspended material all have different size distributions, as illustrated in the example of the Amazon River (Figure 10.1). It should therefore be kept in mind that different size materials have different compositions. Superimposed upon the inorganic fraction of the suspended material transported by rivers is the natural biologically produced material in the river—generally amounting to only a few milligrams per liter of organic material. For the most part, the portion of the suspended material in rivers and estuaries that is traceable to man's activities has a composition similar to the natural material already noted; man has simply accelerated its "erosion." In addition to contributing natural materials to rivers and estuaries, man is responsible for permitting such suspended matter as solid sewage and industrial wastes in the form of, for example, metals, plastics, and woods to enter the waters. In many places this material has been concentrated through natural processes to become a pollution hazard.

TRANSPORT PROCESSES

The first parameter to consider in a discussion of mechanisms of transport of suspended material is the distance the river is capable of thrusting suspended material into the ocean. To obtain data for this parameter, the discharge of the river should be measured through the critical interface region of its estuary or channel. The cross-sectional area of the critical interface region for this calculation may be taken at the contour line of the 1 ppt salinity at the surface; or if the 1 ppt contour extends into the ocean, the cross section should be measured at the mouth of the estuary. By dividing the river discharge by the area of the cross section, a "flushing" velocity may be obtained. This flushing velocity is a measure of the river's ability to keep seawater from intruding into the estuary or, conversely, its ability to thrust its water and materials through the estuary into the ocean. The extent that this flushing veloc-

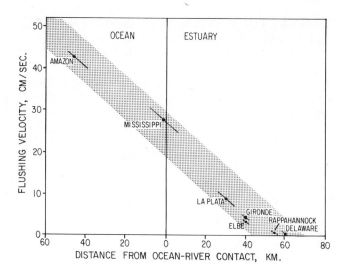

FIGURES 10.2 The position of the turbidity maximum relative to the mouth of various estuaries. Flushing velocity is the mean annual discharge divided by the cross-sectional area at the turbidity maximum.

ity can predict the position of the region of interaction between the river water and its materials and the seawater is illustrated in Figure 10.2. Data from representative rivers of the world show good correlation between the flushing velocity and the position of the river material–seawater interaction. The flushing velocity, therefore, dictates the position from which the mixing and transport mechanisms (waves, tides, and currents) can begin to disperse the suspended material. A general trend indicates that the larger rivers of the world can thrust water and material outward into the ocean, while the smaller rivers do not have the capability of moving material any further oceanward than the upper part of their estuaries. However, it should be noted that, because of the wide range of the areas of the cross sections, many minor exceptions to this overall trend can occur.

The seasonal change in the discharge volume of a river and its corresponding shift in the position of the river-ocean interaction area is represented in Figure 10.2 by the length of the line associated with each of the various rivers. The flushing velocity appears to be the major mechanism controlling the position of the river-ocean interface.

Many estuaries that exist at the mouths of rivers today are not in equilibrium with the present environment; they are the products of the flooding of river valleys by the last major rise in sea level. The estuaries of the rivers that discharged significant quantities of sediment were filled in by this sediment, leaving only their present channels to the sea, e.g., the Mississippi, Amazon, Indus, and Ganges Rivers. At the other extreme, the estuaries of rivers having low sediment discharges, such as the Delaware River and the rivers draining into the Chesapeake Bay, are filling slowly. In time, as these estuaries fill, decreasing the cross-sectional area, the flushing velocity will increase. The increase in

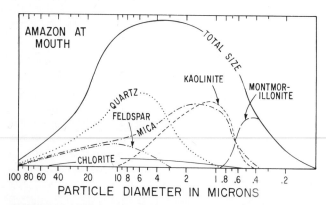

FIGURE 10.1 Size and mineralogic frequency distribution of the suspended solids of the Amazon River. From Gibbs (1967).

the flushing velocity will cause an obvious seaward shift in the position of the river–ocean interaction with time.

Above the estuary, the flow of a river is restricted to a channel, resulting in reasonably high velocities and turbulence. The ability of the river channel to transport sediment is generally more efficient than are the transporting mechanisms in the estuary or the ocean. The river channels of most large rivers have sufficient turbulence to counteract the downward settling of many particle sizes. The 2-μm particles carried are uniformly distributed throughout the water column; the 20-μm particles (silt) show a minor increase in concentration in the bottom fourth of the water column; the 200-μm particles (fine sand) are transported mainly in the bottom fourth of the column, and the 2000-μm particles (coarse sand) are transported by rolling and bouncing along the bottom. As all this material is carried out into the estuary, the river flow is superimposed on a complex and time-variable flow field.

The point to be emphasized is that while the suspended material is transported by the water, it does *not* remain with the same water mass; rather, the particles are continually settling downward, with the finer particles tending to flocculate, enhancing their settling velocity. For studies of suspended-material transport, information on the physical motion of the water is therefore exceedingly useful for consideration with suspended-material data. Data on either aspect alone are of limited value in understanding the transport of the suspended material.

To illustrate the transportation and fate of suspended material delivered to an estuary, Figure 10.3 shows the average flow conditions encountered by the particles of the Amazon River as they are discharged. As the flow leaves the river channel and enters the larger cross-sectional area of the estuary, the flow velocity generally decreases. The particles of larger sizes can settle to the bottom, in many cases before any appreciable seawater is encountered. The material so deposited is generally all the material larger than about 60 μm in diameter, only a small percentage (2–10 percent, generally) of the total load delivered to the estuary (Figure 10.1). As the freshwater, with its suspended material, progresses seaward, it begins to encounter the denser seawater at the critical interface region of river–

ocean interaction. Figure 10.3 shows the less dense freshwater riding up over the seawater as it flows seaward. To replace the seawater entrained and carried seaward in the upper layer, a compensating landward flow is developed in the lower layer (Figure 10.3).

As the particles settle out of the upper layer into the lower layer, they will be carried landward and concentrated to form the turbidity maximum. The particles not only settle as individual particles; but as the seawater is entrained in the upper layer it becomes brackish, causing the finer particles to flocculate, enhancing their settling into the lower layer.

In general, an equilibrium is established between river input and the dispersion of sediment from the turbidity maximum through the transport agents of waves, tidal currents, and nontidal currents. In a simple model of this system, we can see that, once equilibrium has been established, the river input, or the total sediment discharged, must be balanced by transport out of the turbidity maximum and subsequent deposition in one of the environments. The flushing velocity, defined earlier, controls both the position of the river–ocean interaction and the position of the turbidity maximum, thereby actually controlling which of the transportation agents the material will be exposed to. In the case of an estuary having a low flushing velocity, the turbidity maximum will develop in the upper part of the estuary and never be exposed to the oceanic transporting agents.

To elucidate some of the principles that have been presented, let us examine the Amazon River–Atlantic Ocean system. As noted in Figure 10.2, the flushing velocity of the Amazon River is capable of thrusting its water and materials out into the ocean, where a turbidity maximum forms, as shown in Figures 10.4 and 10.5. The water circulation at the 20 percent isohaline shown in Figure 10.3 is substantially the same as that depicted in the model for the turbidity maximum. This two-layer flow extends for over 1000 km northwest of the mouth of the Amazon River (Figure 10.6). Near the river mouth, the driving force causing the landward bottom flow is the density gradient flow, but farther to the northwest it is an Ekman-type upwelling that provides the landward flowing bottom water. These bottom

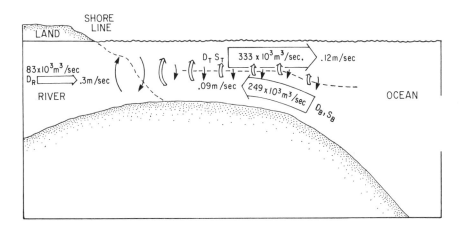

FIGURE 10.3 Model of the circulation off the Amazon River calculated at the 20 ppt isohaline. From Gibbs (1970).

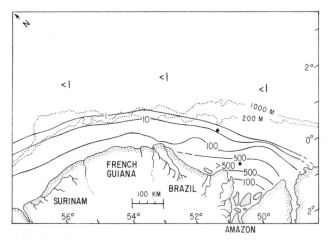

FIGURE 10.4 The near-bottom distribution of the suspended solids in mg/liter for June–July 1971. Dots show positions of anchored stations.

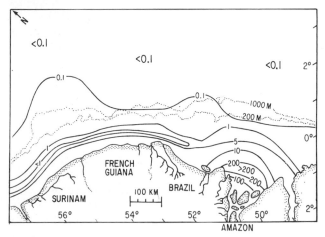

FIGURE 10.5 The near-surface distribution of the suspended solids in mg/liter for June–July 1971.

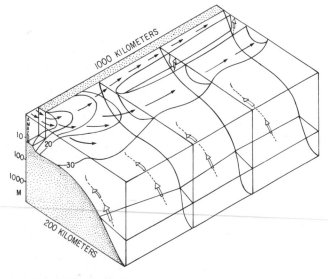

FIGURE 10.6 Three-dimensional model of the water circulation and salinity for the Amazon River–Atlantic Ocean system.

flowing currents have been inferred from salinity and temperature data and also from direct measurements at 14 sites (Gibbs, 1976). The volume of seawater flowing landward along the bottom is approximately three times the river discharge. To verify further this model of suspended material transport, Figures 10.7 and 10.8 present data from two anchored stations in which current direction and velocity and suspended-material concentration were obtained at numerous depths through a complete tidal cycle. Figure 10.7 represents a station near the mouth of the Amazon River and is on the landward side of the turbidity maximum. While the direction of sediment transport varies within the tidal cycle, it should be noted that the net transport during a tidal cycle is offshore, with the bulk of the material moving in the bottom half of the water column. For comparison and to elucidate the mechanisms that maintain the turbidity maximum, Figure 10.8 presents data for a station farther out on the shelf. In this case, the landward transport of suspended material along the bottom is evident. It is this bottom transport that maintains the turbidity maximum on the seaward side. The seaward transport in the middle third of the water column approximately balances the shoreward transport along the bottom. Much of the material that is carried seaward in this middle third probably later settles into the lower layer and moves shoreward again. It should be emphasized that insignificant transport takes place in the upper third of the water column.

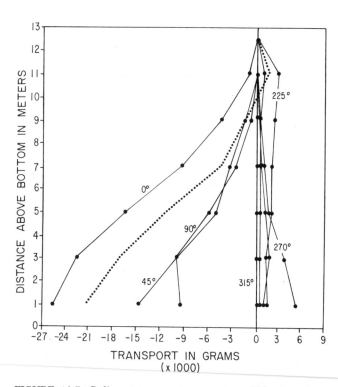

FIGURE 10.7 Sediment transport per meter width at various depths and directions in the ocean oceanward of the Amazon River. The dotted line is the summation computed over a tidal cycle for the onshore-offshore component. The position of the station is shown in Figure 10.4.

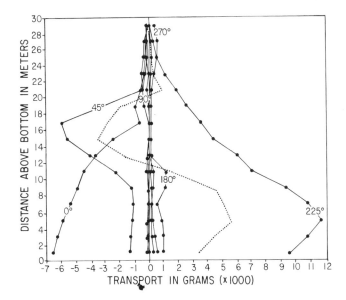

FIGURE 10.8 Sediment transport per meter width at various depths and directions on the outer shelf seaward of and shown in Figure 10.7. The dotted line is the summation computed over a tidal cycle for the onshore-offshore component. The position of the station is shown in Figure 10.4.

FUTURE RESEARCH

Since the particulate materials are a major pollutant transfer agent in our environment and are the materials with which nature fills man-made harbors and channels, an understanding of their mechanisms of transportation is important. We do not understand adequately the processes themselves nor the environmental conditions that are conducive to suspended-material transportation.

Detailed studies of the interaction of the suspended particulate material carried by rivers as it enters seawater are necessary. These studies are needed to ascertain the micromechanisms involved in the overall transportation process. Such topics include the state of flocculation in the natural environment under various current and turbulence conditions and the interaction of the biota considering both the effects of algae and filter-feeding organisms.

An integrated study encompassing the physical movement of the water, the details of the suspended-material transport, the chemistry of the interacting waters, and the biota that may have an effect on the suspended material could elucidate more knowledge than an uncoordinated study of the four separate parts. These integrated studies must entail physical, chemical, geological, and biological disciplines that are truly integrated from the planning stages through interpretation phases.

To comprehend completely transportation processes related to suspended material, a wide range of river–ocean environment systems must be studied. All too often a particular estuary is studied because it is conveniently located rather than because it possesses a natural experiment that will help to solve a fundamental problem. Each new study should address this problem by asking what the unsolved fundamental problem is and whether the proposed study in the particular estuary can be expected to provide a solution.

REFERENCES

Conomos, T. J., and D. H. Peterson (1974). Biological and chemical properties of the San Francisco Bay turbidity maximum, *Mem. Inst. Geol. du Bassin d'Aquitaine 7*, 45.

Cronin, L. E., ed. (1975). *Estuarine Research,* Vols. I and II, Academic Press, New York.

Garrels, R. M., and F. T. Mackenzie (1971). *Evolution of Sedimentary Rocks*, W. W. Norton & Co., New York.

Gibbs, R. J. (1967). The geochemistry of the Amazon River system: Part I. The factors that control the salinity and the composition and concentration of the suspended solids, *Bull. Geol. Soc. Am. 78*, 1203.

Gibbs, R. J. (1970). Circulation in the Amazon River Estuary and adjacent Atlantic Ocean, *J. Marine Res. 28*, 113.

Gibbs, R. J. (1974). *Suspended Solids in Water*, Plenum Press, New York.

Gibbs, R. J. (1976). Amazon River sediment transport in the Atlantic Ocean, *Geology 4*, 45.

Holeman, J. N. (1968). The sediment yield of major rivers of the world, *Water Resources Res. 4*, 737.

Krone, R. B. (1962). Flume studies of the transport of sediment in estuarial shoaling processes, Hydraulic Engineering Laboratory and Sanitary Engineering Research Laboratory Report, U. of California, Berkely, Calif.

Krone, R. B. (1966). Predicted suspended sediment inflow in the San Francisco Bay system, Report to Central Pacific River Basins Comprehensive Water Pollution Control Project, Federal Water Pollution Control Project, Federal Water Pollution Control Administration (Southwest Region), Washington, D.C.

Meade, R. H. (1972). Transport and deposition in estuaries, in Environmental framework of coastal plain estuaries, *Geol. Soc. Am. Mem. 133*, pp. 91–120.

Nelson, B. W. (1973). Environmental framework of coastal plain estuaries, *Geol. Soc. Am. Mem. 133*.

Nichols, M. M. (1972). Sediments of the James River Estuary, Virginia, in Environmental framework of coastal plain estuaries, *Geol. Soc. Am. Mem. 133*, pp. 169–212.

Nichols, M. M., and G. Thompson (1973). Development of the turbidity maximum in a coastal plain estuary, final report to the U.S. Army Research Office, Durham, N.C.

Peterson, D. H., T. J. Conomos, W. W. Broenkow, and P. C. Doherty (1975). Location of the nontidal current null zone in northern San Francisco Bay, *Estuarine Coastal Marine Sci. 5*, 1.

Schubel, J. R. (1968). Turbidity maximum of northern Chesapeake Bay, *Science 161*, 1013.

Schubel, J. R. (1972). Distribution and transportation of suspended sediment in Upper Chesapeake Bay, in Environmental framework of coastal plain estuaries, *Geol. Soc. Am. Mem. 133*, pp. 151–167.

Schubel, J. R. (1974). Effects of tropical storm Agnes on the suspended solids of the northern Chesapeake Bay, in *Suspended Solids in Water*, Plenum Press, New York, pp. 113–132.

The Physical Characteristics and Environmental Significance of Fine-Sediment Suspensions in Estuaries

11

ROBERT R. KIRBY and W. R. PARKER
Institute of Oceanographic Sciences, Taunton, England

INTRODUCTION

The large proportion of the major estuaries in the United Kingdom and Europe are either muddy throughout their length or have dominantly muddy inner reaches. They have a muddy subtidal bed, are bordered by muddy tidal flats, and carry a variable and often high suspended fine-sediment load. The estuarine residual water circulation tends to ensure that fine sediment is retained to form the estuarine turbidity maximum zone. Fine sediments in estuaries are transported in suspension; and although it is not certain, it is widely assumed that the residual circulation of the fine sediment is similar to the residual circulation of the estuarine water.

Fine sediments in estuaries are multimineralic, chemically complex, and may have a significant microbiological population. They consist chiefly of clay minerals of less than 5-μm diameter, which possess significant surface electrochemical charges, resulting in forces that cause them to flocculate. These forces are additional to the normal gravitational and hydrodynamic forces that operate on noncohesive sediments, such as sand; and they cause mud to behave in a fundamentally different manner. As a consequence of these and other problems, knowledge of the behavior of fine sediment lags behind that of noncohesive sediments. Such work as has been undertaken consists largely of laboratory simulations and model tests. However, scaling and other difficulties cause major problems in physical models and other types of laboratory experiments using fine sediments. Furthermore, as is to be discussed later, the physicochemical, physical, and biological processes that operate within estuarine fine-sediment suspensions result in the parameters controlling their behavior being strongly time-dependent.

Present behavioral models of suspensions are based on concepts developed from theoretical, laboratory, and field work. Most models consist of four distinct phases of behavior: erosion, transport, deposition, and consolidation. During the transport phase the models generally assume the movement of undifferentiated suspensions in a turbulent flow, with the flocs remaining in suspension until a slack water period; at such time the flocs are assumed to settle at their still-water settling velocity (usually < 0.1 cm/sec). The amount of settling sediment that reaches the bed depends, of course, on the length of time of the assumed slack water period.

Various authors have attempted to predict instantaneous

suspended sediment concentration profiles. It is not possible to describe theoretically the vertical distribution of sediment in anything other than idealized conditions. The equations due to Rouse, Hunt, Einstein, and Chien are well known (Task Committee on the Preparation of a Sedimentation Manual, 1963). The simplest and most widely used relationship is that of Vanoni (1941) as modified by Einstein (1950). All these formulas predict a concentration profile that increases exponentially with depth, and all relate to non-cohesive sediment. The concentration profiles predicted from these theoretical considerations do differ from those measured in laboratory experiments under equilibrium conditions of steady flow. The differences are generally attributed either to the finite volume occupied by the particles or to their effects on turbulence.

Nordin (1963) as well as others have applied the equations of Vanoni and Einstein to sediment concentration in rivers. Halliwell and O'Connor (1966) attempted to apply these formulas to fine-sediment suspensions in an estuary and also derived modified equations. The profiles calculated using both equations are plotted in Figure 11.1. They attempted to test these equations with field observations but did not have the facility to measure instantaneous vertical turbidity profiles. The time-average profiles they compiled show some results that resemble the predicted distribution but others that depart from it in a fundamental manner. Odd and Owen (1972) divided the exponential profile into two layers as a basis for a model to predict the circulation

of fine sediment in the Thames. A subsequent three-layer model was developed to account for situations where a static suspension formed intermittently on the bed during neap tides (Hydraulics Research, 1974). Linear variations in both velocity and concentration with depth were assumed for the intermediate layer, while the concentration and velocity in the top layer were varied in time but assumed constant with depth.

By comparison with these examples of physical and mathematical modeling, only limited attention has been given to investigations of the behavior of fine sediment in the field or to the development of instrumentation and techniques adequate for performing experiments of equal precision to those undertaken in the laboratory. Thus, the discrepancies between available field data and theory are partially accounted for by the complexity of the field prototype and partially by the inadequacy of the field techniques. The research project at the Institute of Oceanographic Sciences (IOS), Taunton, England, commenced in 1970, has concentrated on a field study of fine-sediment behavior in estuaries. The study has compared the behavior of fine sediment in suspension in two contrasting types of estuary; one is a high-energy, high-turbidity, well-mixed estuary—the Severn—and the other a low-energy, generally very low-turbidity, highly stratified estuary—the Maas.

FIELD TECHNIQUES

At an early stage in the study it became apparent that the gross turbidity structure of the Severn Estuary exhibited important and at that time unpredictable variability. It was therefore our practice to record the horizontal turbidity traverses on a continuous basis. Observations of near-surface turbidity highs associated with bathymetric rises rapidly led us to believe that there were significant vertical turbidity variations. This was confirmed by continuously recorded vertical profiles of turbidity. Use of the simple combined horizontal towing and vertical profiling rig confirmed our suspicion that discrete-point measurements might produce inadequate descriptions of the suspension structures that exist.

The turbidity array designed to investigate these mobile, relatively dilute suspensions consists of a tow-fish upon which were mounted two optical turbidity meters operating in the ranges 250–2000 mg/liter and 500–25,000 mg/liter, a continuous recording conductivity meter, and a pressure sensor. Continuous horizontal traverses were made at towing speeds up to 7 knots. The turbidity meters have a rapid response time (100 Hz), which permitted vertical profiles in depths up to 30 m to be taken in approximately 20 sec. A network of standard stations has been revisited at various stages of the semidiurnal tidal period and at various times in the spring–neap cycle.

To measure denser suspensions, a gamma-ray densimeter was used (Figure 11.2). This transmission gauge was built in 1972 to IOS specification by the United Kingdom Atomic Energy Research Establishment, Harwell. It uses a 3 mCi

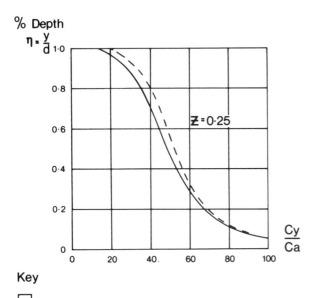

Key

Curve for idealized conditions using data of Vanoni later modified by Einstein.

Curve of Halliwell & O'Connor modified from Vanoni & Einstein to fit an estuarine condition.

FIGURE 11.1 Theoretical vertical suspended sediment profiles. From Halliwell and O'Connor (1966).

FIGURE 11.2   Harwell twin-probe transmissions gauge. The single probe unit is a backscatter gauge used for bulk density measurements.

[133]Ba source and has a vertical resolution of ±1 cm with an accuracy of ±2 percent in the range 1.00–2.00 g/cm³ (Parker *et al.*, 1975). The densimeter can be connected to the same recorders and depth sensor as the turbidity array to provide continuous vertical traverses of density.

Using the turbidity array for relatively dilute suspensions and the densimeter for denser suspensions, we have observed that fine sediment occurs in a continuum of concentrations from dilute, mobile suspensions of only a few milligrams/liter, through "soupy" mobile suspensions with a concentration of 25,000–30,000 mg/liter, to dense static "custardlike" suspensions of 50,000–400,000 mg/liter.

FINE-SEDIMENT PHENOMENA IN THE SEVERN

The Severn Estuary extends from Upton-on-Severn to Flat Holm and Steep Holme Islands, a distance of 90 miles (150 km), (Figure 11.3). There is a large tidal range with large differences in energy levels between spring and neap tides (Table 11.1). The Severn is regarded as a classic, well-mixed estuary, although recent work has shown that marked lateral partitioning occurs. It is a high-turbidity estuary, the turbidity maximum occupying the whole estuary from the Holmes Islands to well beyond the Severn Bridge (Figure

TABLE 11.1   Tidal Data for the Severn Estuary

Tidal Range at Avonmouth (m)		Tidal Stream Velocity at Avonmouth (m/sec)	
		Flood	Ebb
Extreme spring range	14.5		
Mean spring range	12.3	Springs	2.3	1.6
Mean neap range	6.5	Neaps	1.2	0.85

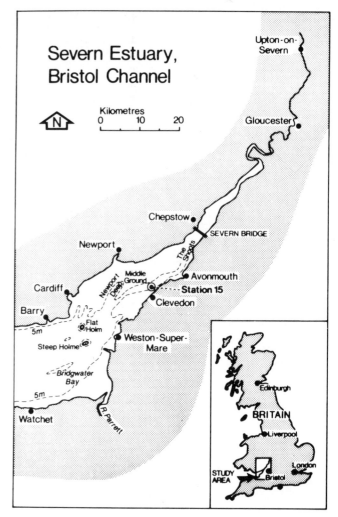

FIGURE 11.3   The Severn Estuary and Upper Bristol Channel, U.K., showing position of Station 15.

11.3). In the area between The Shoots and Bridgewater Bay, of the order of 10^{10} kg of sediment are in mobile suspension during spring tides (Kirby and Parker, 1975).

Fine sediment occurs in two forms: as suspensions, which may be either mobile or static, and as settled mud, which is quasi-permanently deposited on the seabed. Over both the semidiurnal and the spring to neap tidal time scales, predictable sequences of gross concentration profiles have now been established over concentrations ranging from 250 to 400,000 mg/liter. Dense suspensions have been recorded along the whole length of the channels from The Shoots to Bridgewater Bay, but they rarely occur over the banks.

The experiments in the Severn were undertaken using a single ship, which was brought onto station and then allowed to drift with the tide to prevent disturbance of the near-surface suspension structures by either the anchor cable or the propeller. Consequently, the only tidal stream data that could be obtained simultaneously were measurements taken with the ship's EM log.

MOBILE SUSPENSIONS, SEMIDIURNAL CYCLE

Over a single tidal cycle on spring tides a regular sequence of suspension structures develops in the water column throughout the estuary. The sequence commences with a vertically homogeneous profile at maximum ebb or flood and is followed by the development of layering within the suspension during the decreasing tidal energy stage toward and during slack water. On the increasing tidal energy stage following slack water, a further sequence of layered structures is developed.

This general sequence is illustrated in Figure 11.4, which shows the succession of structures measured during a 3½-hour experiment encompassing the low water of a spring tide at Standard Station 15. At the maximum ebb, the turbidity was constant from top to bottom with a concentration of 10,000 mg/liter. At 1356, an inflection in the concentration profile was observed near the surface. From 1356 to 1441, as the tidal velocity continued to decrease, the inflection within the turbidity profile became less pronounced and occurred deeper in the water column, while the turbidity of the lower layers increased progressively. Profiles at 1443 and 1446 are almost logarithmic and show an increased concentration in the lower part of the profile. While this sequence was developing, the top 4 m of the water column was moving down the estuary at approximately 1.0 m/sec. By 1504, the time of predicted low water, a dense, presumed mobile layer 0.25 m in thickness and with a concentration approaching 40,000 mg/liter was observed at the seabed. The large-amplitude turbidity change across the surface of this layer represents a density change from approximately 1.025 to 1.055 g/cm³. From 1528 on, the tide at the surface was flooding and the water depth started to increase. However, despite this 180° change in tidal current direction at the surface, between 1504 and 1550 the continuing decrease in the altitude of the boundary between a relatively clean (1000 mg/liter), constant-turbidity upper layer and a higher-turbidity (approximately 10,000 mg/liter) layer below it can clearly be seen. During this period, the thickness of the lowest layer increased, until by 1550 it had reached 2 m. By 1558, the high-level step in the turbidity profile had disappeared and the sediment concentration in the upper part of the profile had decreased to 1000 mg/liter. The lowest layer had reached 3 m in thickness, and the surface flood tidal current was 1.25 m/sec. The succeeding profiles appear to show the sediment rising progressively into the water column, the lowest layer in the profile becoming thinner and apparently less concentrated. At 1630, the surface flood velocity had reached 1.75 m/sec, and by the end of the experiment at 1730 a homogeneous profile almost identical with the first one recorded at 1356 was observed. Comparison of data from other standard stations shows that the sequence of events detailed from Station 15 may be anticipated in any of the deep channels under similar conditions at high or low slack water. Stepped profiles are almost universally present during this sequence.

Results of this type of time series observation are believed to show two major effects. The inflections in the turbidity profiles marking abrupt vertical changes in sediment concentration are believed to be the product of settling of individual flocs. The movements of these inflections up and down the profile are interpreted as being partially due to advection past the measuring point of interfaces generated upflow, since their vertical movements are apparently too rapid to equate with the maximum expectable settling velocities of the flocs. The fact that the inflections appear to move up and down in phase with the tidal currents suggests that they are inclined. Interpretation of the data for Station 15 suggests that the highest interface on the records taken from 1504 to 1556 slopes at approximately 1:100.

MOBILE SUSPENSIONS, SPRING-NEAP CYCLE

On the second timescale, the spring-to-neap cycle, the semidiurnal sequence of homogeneous, followed by stepped,

STATION 15   2-8-74   BC 1974/3   F.S.D 0-5

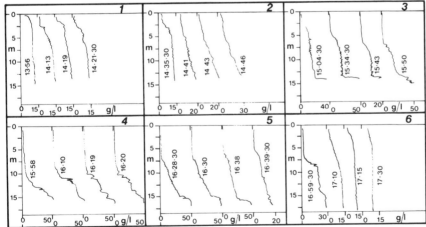

TURBIDITY/DEPTH PROFILES

FIGURE 11.4   Time series of vertical turbidity profiles from Station 15, Severn Estuary.

suspensions continues. However, as the tidal range and consequent tidal stream velocities decrease, dense, mobile suspensions differentiated onto the bed become static at slack water. These static suspensions survive progressively longer into the succeeding tide, until they eventually resist the subsequent elevation in tidal energy and remain throughout the complete tidal cycle. At this stage of the neap cycle, the concentrations in the overlying water body are greatly reduced and a recognizable sequence of structures may not exist. In general terms, the mobile suspension structures on neap tides resemble those at slack water on a spring tide in that the concentrations in the overlying water column are much reduced, and most of the sediment is on the bed as a dense static suspension.

STATIC SUSPENSIONS

Dense static suspensions, commonly 2–4 m and sometimes reaching 10 m in thickness, occur in the channels of the Severn and in Bridgewater Bay during neap tidal times. Soon after becoming static, the upper surface of the suspension becomes sufficiently well defined to be detected with an echo sounder (Figure 11.5). Static suspensions with concentrations in the upper layers as low as 14,000 mg/liter as measured with turbidity meters have been detected with a 30-kHz echo sounder. Densities ranging between 1.05 and 1.3 g/cm³ as measured with the density gauge are regularly encountered on neaps. Such dense static suspensions are a common feature in estuaries, being called in the United Kingdom "fluid mud" and in the United States "fluff" by hydrographic surveyors.

Examination of dense suspensions using echo sounders reveals that detection of the suspension is dependent on both the frequency of the acoustic source and the surface character of the suspension. Figure 11.6 shows records obtained simultaneously using a 200- and a 30-kHz echo sounder over a static suspension. The characteristics of the interface are such that it was detected by the 200 but not by the 30-kHz source. The relative importance of back-scattering compared to reflection from the interface is

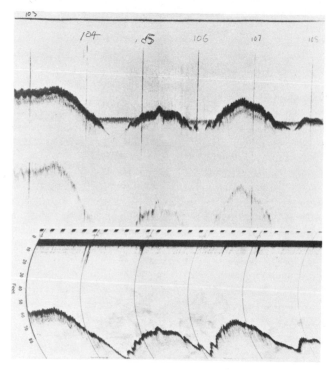

FIGURE 11.6 Comparison of 200-kHz and 30-kHz records obtained simultaneously, Severn Estuary. The upper record is at 200 kHz and the lower at 30 kHz.

unknown. Knowledge of the distribution of static suspensions in the Severn was gained using the lower-resolution 30-kHz source, and thus our estimate of the abundance of such dense static suspensions is conservative.

In some localities, multilayered static suspensions occur, the layering being generally parallel to the upper surface of the suspension rather than the underlying seabed. Continuous vertical profiles of density measured with the transmission gauge show that even apparently uniform suspensions identified with the 30-kHz echo sounder may be stratified. Figure 11.7 shows two density profiles into static

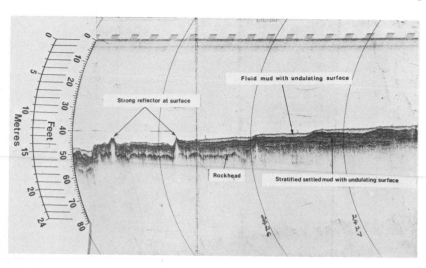

FIGURE 11.5 30-kHz echo-sounder record showing dense static suspension, Newport Deep, Severn Estuary.

suspensions near Avonmouth. The lower profile shows layering visible on the echo-sounder record, which is recognizable as distinct steps in the density curve. The upper profile shows a stepped density profile in an acoustically undifferentiated suspension.

In some circumstances, acoustic records of static suspensions show reflectors near the base of the suspension, which are parallel to the underlying topography rather than parallel to the planar surface of the suspension (Figure 11.8). By analogy with structures in settled muds, this is believed to indicate a stage in the transition from static suspensions to settled mud. Extensive deposits of settled mud (Figure 11.9) having a surface density >1.3 g/cm^3 occur within the estuary.

During the neap cycle the suspensions continue to dewater, but later, when tidal energy levels start to rise on the succeeding neap-to-spring cycle, a large proportion of all static suspensions is eroded and remixed up into the water column.

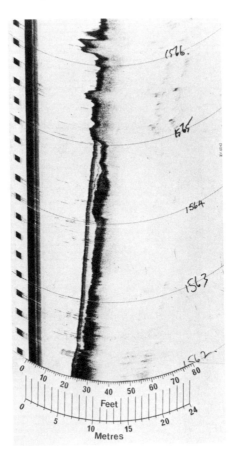

FIGURE 11.8 Dense static suspension in which lower layer is at a transitional stage of dewatering to form settled mud, Severn Estuary.

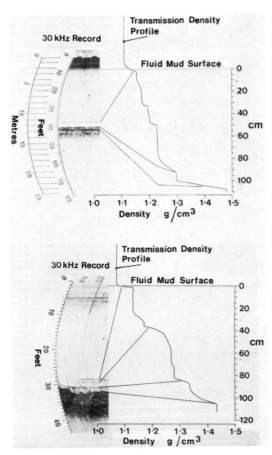

FIGURE 11.7 Comparison of transmission gauge profiles into dense static suspensions. Accompanying 30-kHz echo sounder records obtained at time of profiles, Severn Estuary. The density discontinuities on the lower record are also observed acoustically; those on the upper record are not.

FINE-SEDIMENT PHENOMENA IN THE MAAS

Measurements using turbidity meter and densimeter arrays were made in the Maas for comparison with data from the Severn. The Maas, a distributary of the Rhine and site of the port of Rotterdam–Europoort, is a highly stratified, very low-tidal-range (1.4 m) and generally low-turbidity estuary. It might thus be expected that contrasts in fine-sediment phenomena might be seen. A 13½-hour tidal stream and turbidity survey was undertaken using three vessels along a section crossing both the Rotterdam Waterway and the Caland Canal, which is the entrance to Europoort (Figure 11.10). Vessels were anchored for the tidal stream observations at positions 2 and 4, while a third sailed repeatedly across the section recording vertical profiles of turbidity and conductivity at five positions. Only significant parts of the 13½-hour experiment are plotted in Figures 11.11 and 11.12.

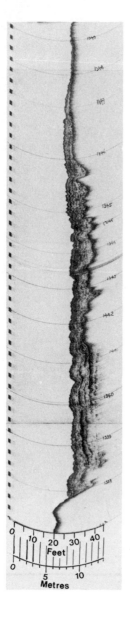

FIGURE 11.9 Settled mud, Bridge-water Bay, Upper Bristol Channel.

FIGURE 11.10 The Maas Estuary, Netherlands, showing position of experimental cross section.

MOBILE SUSPENSIONS

The results show the generally low turbidity of the estuary. For much of the time, the turbidity was probably in the region of 20–40 mg/liter, below the resolution of the turbidity meters used at that time; but dense mobile suspensions were detected at both center-line stations. Results from Station 2 (Figure 11.11) show a salt wedge, a reversing tidal flow, and low turbidity. The flow at the surface was downstream throughout the experiment, while flow at the bed was upstream except for a period of less than 2 hours toward the end of the ebb. Surface velocities varied between 100 and 200 cm/sec. The turbidity values recorded close to the anchored station by the suspended solids-profiling vessel show low concentrations throughout the experiment except for one profile 1 hour before high water, which shows a near-bed dense suspension 0.5 m in thickness with a concentration of 40,000 mg/liter. Comparison with the current data shows that this suspension was almost certainly moving upstream at 25–30 cm/sec.

At Station 4 in the Caland Canal, the salt wedge was only poorly developed because of the limited freshwater flow. The tidal stream velocities were lower than those in the Rotterdam Waterway, and the near-bed flow was upstream at all stages of the tide (Figure 11.12). At the surface, the flow was upstream for the 3 hours leading up to the high water. Just before high water, the surface current reversed and continued to flow downstream for the remaining 9 hours of the experiment. The near-bed upstream velocity reached 30 cm/sec on the flood and fell to 10 cm/sec during the maximum ebb. At the surface, the flood velocity reached 30 cm/sec, while the ebb velocity reached 75 cm/sec. The turbidity records from Station 4 reveal that, with the exception of the occasional profile showing sediment mixed high into the water column by the passage of working dredgers, the turbidity was much less than 200 mg/liter. However, at 1205 hours, in the late part of the ebb, a mobile suspension

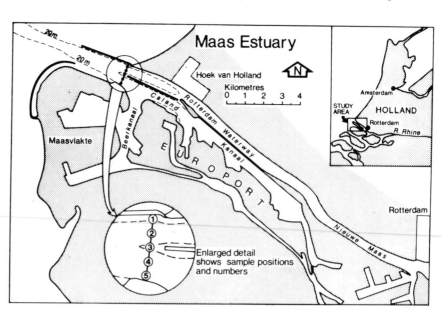

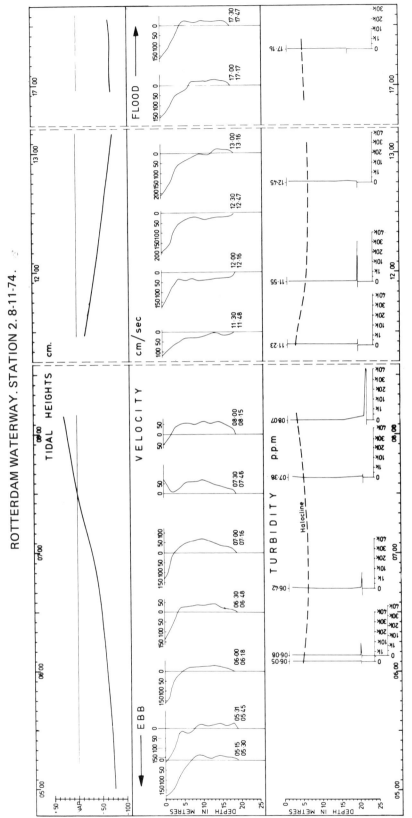

FIGURE 11.11   Plots of tide height, velocity, and direction and turbidity for Station 2, Mass Estuary.

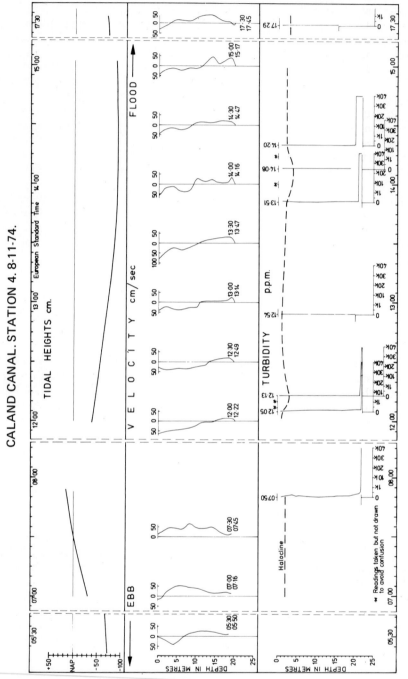

FIGURE 11.12 Plots of tide height, velocity, and direction and turbidity for Station 4, Mass Estuary.

was observed as a layer 0.25 m in thickness with a concentration reaching 40,000 mg/liter. The profiling rate was increased, and during the next 2½ hours the layer was observed to increase progressively in thickness until by 1420 hours it was 1.75 m thick. During this time the near-bed current was moving upstream at 15–25 cm/sec. The concentration was at the upper end of the resolution of the turbidity meter so that no internal structure could be detected, but the conductivity profiles obtained simultaneously with the turbidity records revealed that the dense suspension was layered.

There is no evidence of the source of this dense suspension. It is apparent, however, that all the significant suspended sediment transport took place in dense near-bed layers. More significantly, the suspensions were moving upstream, and since there was no return flow at the bed they would be unlikely to escape from the estuary if similar conditions persisted until the suspension had dewatered and consolidated. The largest quantity of sediment penetrated upstream at what is probably the least expected time, toward the end of the ebb tide. Examination of the tidal stream profiles and Rijkswaterstaat monitoring indicates that the dense suspensions are unlikely to have originated upstream of the measuring section in either the Rotterdam Waterway or the Caland Canal.

STATIC SUSPENSIONS

The Caland Canal and the basins of Europoort and Rotterdam are the sites of extensive dense static suspensions. Research undertaken by the Rijkswaterstaat using echo sounders and the Harwell densimeters indicates that static suspensions represent large quantities of sediment, which appear to accumulate rapidly during storm periods. Their work indicates that once the sediment enters the Maas and is deposited it is permanently trapped and progressively dewaters to form settled mud. Such observations agree with the current-meter measurements made during our investigations.

ENVIRONMENTAL SIGNIFICANCE

Estuaries are of extreme importance in the interrelation of man with his environment. They are the sites of almost all the major cities and seaports of the world. In the large proportion of these estuaries, as a consequence of the retentive estuarine sediment circulation, the dominant sediment type is mud. Both dense mobile and static suspensions have a wide environmental significance. The dense mobile suspensions in the Severn and Maas are probably the most important source for cohesive sediment arriving at or returning to sites of accretion in navigational channels, locks, and basins.

Two particular engineering problems related to the presence of fine sediment may be identified. The first is that layered static suspensions are generally surveyed with echo sounders that are incapable of identifying which horizons should be regarded as the navigationally significant seabed. Fortunately, densimeters are capable of resolving certain of the problems related to determining the altitude of the seabed (Kirby and Parker, 1974), and they are now being used extensively in this role by the Rijkswaterstaat. The second problem is that of dredging and disposing of cohesive sediment. It is difficult for the dredger to detect or to dredge dense static suspensions, and the underlying dredgeable material is greatly diluted by the dredging plant in order to move it from the seabed. Dense mobile suspensions are generated by the passage of the draghead while high level suspensions are created by overflow through the weirs of the dredger hopper. At the dumping site, dense suspensions can be formed, which may assist dispersal of the sediment and its return to the dredging site or to other areas of importance such as shellfish beds, fishing grounds, or recreational beaches.

Cohesive sediments are environmentally important in the pollution field. It has been shown (Preston, 1973; Kirby and Parker, 1973) that in turbid estuaries a high proportion of the heavy-metal pollutants are carried as an absorbed load on the surface of the clay particles. Analyses of mean concentrations of heavy metals in the waters of the Bristol Channel showed 1.18 μg/liter of Pb, 9.98 μg/liter of Zn, 2.07 μg/liter of Cu, and 1.13 μg/liter of Cd (Abdullah *et al.*, 1972). In contrast, suspended sediment samples from the Avonmouth area, which show the highest heavy-metal contents in the estuary, range from 140 to 160 ppm (dry weight) Pb, 400 to 570 ppm Zn, 43 to 50 ppm Cu, and 2 to 5.6 ppm Cd. Similar concentrations occur at the upstream end of Newport Deep with a general progressive decrease seaward to lower values in Bridgewater Bay. The mobile fine sediment, whether dispersed or in the form of dense static suspensions, provides a reservoir for heavy-metal pollutants in the estuary because of the retention of fine sediment by the estuarine circulation system. Even if the suspended sediment is merely a carrier for the pollutants rather than the ultimate dump site, the reservoir of pollutants present is large.

In the Severn, the presence of high concentrations of suspended sediment as discrete layers is sufficient to affect the dynamic balance of the estuary. The work on the Maas shows that as a consequence of the formation of density stratification due to fine-sediment concentration differences the circulation of fine sediment may be quite different from the circulation of the water. Thus, the existence of fine sediment and the nature of the fine-sediment circulation are likely to be of particular importance in modeling the dispersal of pollutants.

Dense mobile and static suspensions can have a fundamental effect on the ecology of an estuary. In addition to the pollution load carried by the sediment, the constant development of extremely high-turbidity suspensions may be intolerable for some benthic fauna and flora. In the Severn, the main channel, the site of regular formation of mobile dense suspensions, has a poor benthic fauna largely confined to the sedentary tube-building polychaete worm *Sabellaria* sp., which colonizes the channel sides. The mobile dense suspensions may be largely responsible for this restricted fauna. Similarly, every neap tide dense static suspension completely blankets the settled mud areas with 1–2 m of material at concentrations of 50,000–400,000 mg/liter.

Recent work undertaken in conjunction with Bristol University on the microbiological condition of these static suspensions in the Severn has revealed that they are apparently largely anaerobic. The presence of such an anaerobic blanket, either at regular intervals as in the Severn or more irregularly but longer lived as in the Maas, is likely to be of prime importance as an ecological control mechanism.

CONCLUSIONS

Detailed field investigations using new techniques in two contrasted European estuaries have revealed that the behavior of fine sediment is poorly understood and that laboratory simulations are unlikely to duplicate certain of the significant real-scale and time-dependent effects. The channels of the Severn are the site of mobile dense suspensions and ephemeral static suspensions on springs and more stable dense static suspensions on neaps. A sequence of structures develops in response to changes in energy levels on the semidiurnal and spring–neap time scale. In contrast, in the Maas dense mobile suspensions and static suspensions have been identified whose development has little apparent relation to any tidal time scale. It may be that only a one-way movement of dense near-bed suspensions occurs in the Maas.

Despite the great contrasts in physical character, both estuaries studied show temporally and spatially complex and stratified suspensions, and it may be that similar structures are common in other estuaries. In both estuaries, dense mobile suspensions are thought to play an important role in the sediment circulation.

Instantaneous profiles approximating the usually predicted exponential concentration profiles have only occasionally been observed, although an approximation to an exponential profile would result if continuously measured vertical profiles of concentration were integrated over a sufficiently long period. The complex sequences of structures that actually occur need to be taken into account in any models of suspension behavior.

Dense mobile suspensions might be expected in many high-turbidity estuaries and also in low-turbidity estuaries with mud substrates that are subject to dredging.

Environmental problems due to suspended sediment might be encountered even in estuaries with a high but dispersed suspended load. It is suggested that any dense near-bed mobile or static suspensions are likely to give rise to a wider variety of serious environmental problems.

Future work is being directed toward a quantitative description of fine-sediment phenomena and to field investigations of the processes governing their behavior.

When the phenomena themselves are adequately understood and the processes responsible for them are adequately identified and scaled, the resulting realistic models will form the basis from which predictive models of estuarine sediment circulation and fine-sediment behavior may be developed.

ACKNOWLEDGMENTS

The authors are grateful to the officers and crew of RV Edward Forbes, the vessel used for survey work in the Severn, to the Netherlands Rijkswaterstaat, which provided the facilities and cooperated in the Maas survey, and to Michael Moore for assistance in laboratory and field work. The continued support of the Port of Bristol Authority is gratefully acknowledged. This project is funded by the Departments of Industry and of the Environment. The microbiological analyses were financed by Bristol Corporation.

REFERENCES

Abdullah, M. I., L. G. Royle, and A. W. Morris (1972). Heavy metal concentration in coastal waters, *Nature 235,* 158.

Einstein, H. A. (1950). The bed-load function for sediment transportation in open channel flows, U.S. Dept. of Agriculture Tech. Bull. No. 1026, Washington, D.C., pp. 1–71.

Halliwell, A. R., and B. A. O'Connor (1966). Suspended sediment in a tidal estuary, *Proceedings of the Tenth Conference on Coastal Engineering,* Am. Soc. Civil Eng., pp. 687–706.

Hydraulics Research (1974). The report of the Hydraulics Research Station, Wallingford, U.K., pp. 1–86.

Kirby, R., and W. R. Parker (1973). Fluid mud in the Severn Estuary and Bristol Channel and its relevance to pollution studies, paper presented at a symposium on Estuarine and Coastal Pollution sponsored by the Inst. Chem. Eng.

Kirby, R., and W. R. Parker (1974). Seabed density measurements related to echo sounder records, *Dock and Harbour Authority 54,* 423.

Kirby, R., and W. R. Parker (1975). Sediment dynamics in the Severn Estuary: A background for studies of the effects of a barrage, in *An Environmental Appraisal of the Severn Barrage,* Dept. of Civil Eng., U. of Bristol, Bristol, U.K., pp. 35–46.

Nordin, C. F., Jr. (1963). A preliminary study of sediment transport parameters, Rio Puerco near Bernardo, New Mexico, Professional Paper 462-C, U.S. Geological Survey, Washington, D.C., pp. 1–21.

Odd, N. V. M., and M. W. Owen (1972). A two-layer model of mud transport in the Thames Estuary, *Proc. Inst. Civil Eng. 7517S,* 175.

Parker, W. R., G. C. Sills, and R. E. A. Paske (1975). *In situ* nuclear density measurements in dredging practice and control, *BHRA First Internat. Symp. on Dredging Technol. B3,* pp. 25–42.

Preston, A. (1973). Cadmium in the marine environment of the United Kingdom, *Marine Pollut. Bull. 4,* 105.

Task Committee on the Preparation of a Sedimentation Manual (1963). *Am. Soc. Civil Eng. J. Hydraul. Div. 89,* HY5, 45.

Vanoni, V. A. (1941). Some experiments on the transportation of suspended load, *Trans. Am. Geophys. Union 22,* 608.

The Fate of Metals in Estuaries

12

KARL K. TUREKIAN
Yale University

INTRODUCTION

From the geochemical point of view, an estuary is a reaction vessel where the mixing of stream water and seawater has consequences far beyond simple dilution, for streams bring more than freshwater to the basin. They are bearers of dissolved organic compounds and inorganic species and of detrital material including organic matter and iron and manganese oxide grains and coatings as well as the minerals from soil profiles. And the sea provides the well-known dissolved chemical species that make it salty. Of these, sulfate plays a special role in determining chemical pathways in the estuary, especially in relation to biological activity in the sediment column. In addition, the dynamics of estuarine circulation and the activity of benthic populations force intimate contact between the water column and the sediment pile.

All techniques capable of helping us to understand this complex system are welcome, even when they yield results leading to diametrically opposite conclusions; for it is at the point of contradiction that we can see how much our analysis of the system has been oversimplified or how we have structured too artificial a framework to represent this complex reaction vessel.

One of the most pressing problems in estuarine geochemistry is the behavior of the trace metals. Here the modifications of stream supplies occur, and the quality and quantity of what is actually delivered to the open ocean are determined. One of the best ways to understand the behavior of trace metals in estuaries is to study the behavior of manganese, iron, and the daughters of the uranium (and thorium) decay series nuclides in the stream–estuary–ocean system. These provide the opportunity to determine rate constants in natural systems. The objective is to use these constants, at least in modeling, to describe the behavior of other elements for which data are difficult to obtain directly.

The special properties of manganese and iron are that they readily undergo oxidation and reduction under the conditions available in different parts of an estuarine system and they are generally high in abundance in the sediments. The anoxic sediments in which seawater sulfate in pore waters is reduced by bacteria to form hydrogen sulfides also reduces iron and manganese in the sediment to Fe^{+2} and Mn^{+2}. As the concentrations of these ions are determined by the solubilities of their sulfides, their concentrations are quite high in pore waters—a property of manganese and iron and not of other common metal sulfides.

121

(Manganese solubilities may also be controlled by the stability of $MnCO_3$.) The release of Mn^{+2} and Fe^{+2} into aerated waters by biological and physical mechanisms results in their oxidation to Fe^{+3} and Mn^{+4}, respectively, and precipitation as oxides. The freshly precipitated iron and manganese oxides provide highly reactive surfaces, which sequester many trace metals.

One of the members of the uranium decay series (Figure 12.1) of interest in modeling the behavior of metals is ^{210}Pb, which has a half-life of 22 years. This nuclide is supplied to aqueous systems in part by the decay of ^{222}Rn mainly derived from ^{226}Ra. The atmosphere is the other source from which the decay of ^{222}Rn, which continuously emanates from soils to the atmosphere, results in a supply of ^{210}Pb, which is transported to ground level by precipitation.

Another nuclide in the uranium decay series that is of value in predicting the behavior of trace metals in estuarine waters is ^{234}Th. It is supplied to estuarine waters exclusively by the decay of dissolved ^{238}U. This, and the fact that its half-life is 24 days, provides direct information on the systematics of the metal scavenging process in an estuary.

The following are some of the insights that have been gained on these systems by our group at Yale University. In addition, some of our work on the stable trace metals is interpreted in light of experiments recently performed by others.

THE SUPPLY OF METALS BY RIVERS

The composition of nonimpacted streams is the result of the weathering process. Except at times of extremely high rainfall, the water in the stream channels is derived from groundwater. The reactions at depth in the country rock fix the composition of the groundwater supplied to the stream channel. Aerated rain water percolating through the soil loses its oxygen progressively as a result of metabolism by soil bacteria. The major reservoir of groundwater is thus anoxic and provides a reducing environment which influences chemical speciation. The most important effect for trace metal transport is that under reducing conditions manganese and iron can be mobilized. Streams are fed by these manganese- and iron-rich groundwaters and are oxidized and subject to precipitation. Manganese and iron oxide precipitates can sometimes be seen in massive deposits around springs, but most commonly they precipitate on the surfaces of the ubiquitous particles found in a running river.

The deficiency of ^{210}Pb in groundwater generally (Table 12.1), relative to the expected supply either by rain or by production from dissolved ^{222}Rn (mainly supported by ^{226}Ra), implies that absorption of ^{210}Pb has occurred at depth. The trace metal nuclides can be expected generally to be subject to the same sequestering action. After steady state

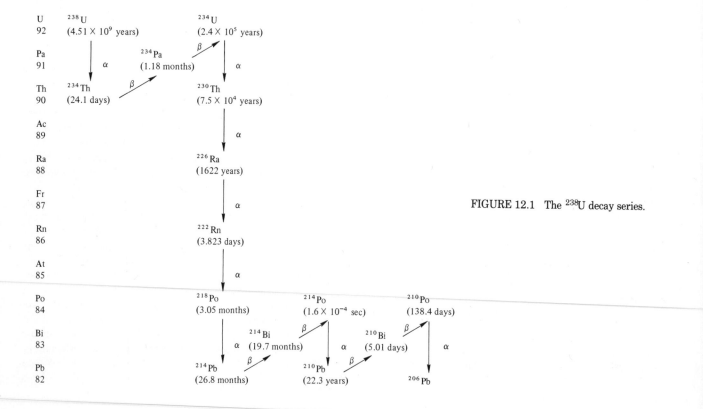

FIGURE 12.1   The ^{238}U decay series.

TABLE 12.1 Summary of Behavior of ^{210}Pb in Ground-water Regimes Based on Data from Holtzman (1964)

Rain ^{210}Pb = 10 dpm/liter Evaporation-Transpiration = $50\frac{1}{2}$	Land Surface
Infiltration ^{210}Pb = 20 dpm/liter	Water Table
^{210}Pb = 0.4 dpm/liter ^{226}Ra = 1 dpm/liter	Shallow Groundwater
^{210}Pb = 0.04 dpm/liter ^{226}Ra = 10 dpm/liter τ = 1 month	Deep Groundwater

between dissolution and absorption is established, however, a measurable flux of metals to the stream may occur, although it may not be seen in the ^{210}Pb concentration. On the basis of ^{210}Pb estimates in groundwater, the mean residence time relative to absorption is less than a month. Some of this removal may be by roots of plants and trees that then transfer ^{210}Pb (as well as stable trace metals) to the surface where they become part of the plant litter. Thus, as rock is disintegrated by the action of soil organisms producing organic acids and carbonic acid, some of the metals are absorbed by the vegetation. Subsequent decay of this vegetal material provides an organic-rich material commonly called top soil. This top soil is the repository of seasons of accumulation of vegetal debris processed and reduced in volume by soil organisms. The residual organic material itself is a strong sequesterer of trace metals, however supplied. Using ^{210}Pb derived from atmospheric precipitation as a tracer, we have shown (Benninger *et al.*, 1975; Benninger, 1976) that virtually every bit of the ^{210}Pb supplied by precipitation is retained by the top soil (Table 12.2).

By the erosion of its banks, a tributary system transports sections of the soil profile including the top soil with its sequestered metals. A material balance calculation of ^{210}Pb in the drainage basin of the West Branch of the Susquehanna River indicates that this nuclide has a mean residence time of about 2000 years in the soil relative to transport by streams (Lewis, 1976).

TABLE 12.2 Atmospheric Flux of ^{210}Pb in the Eastern United States by Direct Measurement and Accumulation in Ground-Level Long-Term Repositories (compiled by Benninger *et al.*, 1975)

Location	Flux (dpm/cm^2/yr)	Type of Measurement
New Haven, Conn.	1	Precipitation and dry fallout
East Haven, Conn.	1	Salt marsh profile
East Haven, Conn.	0.8	Soil profile
Cook Forest St. Park, Pa.	1	Soil profile
Maryland	1.2	Soil profile

As a stream moves its burden of minerals and organic-rich detritus to the sea, it is continuously being fed by groundwaters that supply dissolved materials. However, additional chemical changes can occur in the channel itself. The river bed is an environment in which biological activity can deplete oxygen in pore waters and thus supply soluble manganese and iron to the stream. This is most effective in deeper and wider parts of the stream channel and causes precipitation of oxidized manganese and iron on the suspended material in the stream in a continuous manner down its course.

The act of cycling manganese and iron in the river as well as the supply of additional quantities of these elements in soluble form from groundwater results in an efficient scavenging of trace metals from the stream onto particles. In the case of the natural tracer ^{210}Pb it can be shown that its mean residence time in solution in streams is about one day and seems to be coupled directly to manganese precipitation (Lewis, 1976). Thus, the rate constant for manganese precipitation (and possibly iron) may determine the rate constant for the removal of many trace metals from solution. As manganese is resupplied to the stream from reducing sediments along its course, a fraction of the precipitated metals may also be released, thus providing a steady-state soluble concentration of each of the metals. The low relatively constant concentrations of many trace metals in streams may be explained by this steady-state process.

The trace-element concentrations in particles in streams are related to both the manganese concentration and the organic matter concentration. They are unrelated to the formal ion exchange capacities of the clay minerals typically supplied from weathering profiles. Our early experiments in absorption and desorption of trace metals on clay minerals in freshwater and seawater systems thus are incapable of explaining the major controls on trace-element transport as seen in real streams. It is doutful whether more sophisticated *in vitro* experiments of this kind will provide any new insights into trace metal behavior in natural aqueous systems.

Particles in streams also act to modify the stream composition where artificial injections of soluble trace metals occur (Turekian, 1971). In the Naugatuck River of Connecticut, a tributary of the Housatonic, the Ni, Co, and Ag concentrations drop at least an order of magnitude from the point of injection of industrial metal-rich acid wastes to a point 1 km downstream (Figure 12.2). The metals sequestered on particles are transported to the estuary. A striking example of this is seen in the high-trace-metal concentrations found on suspended particles in the Rhine River as it transects The Netherlands (Martin, 1971).

In summary, we see that the burden of trace metals supplied to an estuary by streams comes primarily on particles. This in turn is related to the organic and manganese concentrations of the particles. By analogy with the behavior of ^{210}Pb in such systems, it is obvious that organic and manganese oxide phases are strong sequesterers of trace metals. What then occurs as this assemblage reaches the sea?

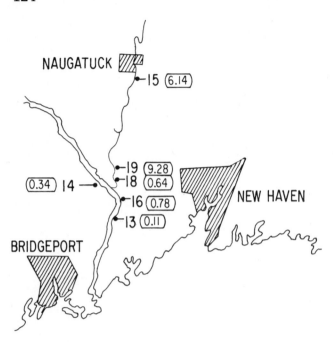

FIGURE 12.2 The distribution of "dissolved" cobalt (μg of Co/liter) in the Naugatuck-Housatonic River system. The Naugatuck River is heavily impaced by industry and high in dissolved metals. The concentration decreases away from the source of impact as the result of absorption on particles.

THE FRESHWATER–SEAWATER ENCOUNTER

The physical boundary where a stream encounters seawater is hardly a simple one. In most places, the action of the tides continuously changes the encounter configuration and any strong change in stream runoff on the one hand or the response of the sea to storms on the other has profound effects on this interface.

There are two diametrically opposite processes involving trace metals that may happen at the freshwater–seawater boundary:

1. The precipitation of iron and manganese and some other elements such as phosphorus, aluminum, and titanium has been shown to occur in both field and laboratory experiments. This is ascribed to the formation and flocculation of colloids as the increased ionic content of the salt water alters charge distributions (Sholkovitz, 1976; Boyle, 1976).

2. The release of metals from particles at the seawater-freshwater interface has also been invoked to explain certain field observations (Kharkar et al., 1968). This process is compatible with laboratory experiments in which trace metals absorbed from freshwater on clay minerals have been shown to release them as the total cationic concentration of the solution increases. This process exists most strikingly as the hydrogen ion concentration increases, but the encounter with seawater does not normally decrease

the pH of solutions. We have seen, moreover, that the trace element burden must be carried by the organic and manganese-rich phases and not by clay minerals.

The field evidence for release of trace metals in the estuarine system is the observation that suspended particles in some European estuaries decrease in trace metal concentrations as the salinity increases (Martin, 1971). The exact method of release is not specified but appears to be the destruction of the metal-bearing phases rather than simple desorption. Such field experiments depend, of course, on the assurance that only the streamborne particles are involved in chemical changes occurring progressively seaward. If nonindigenous sediments are transported into the estuary from the open sea, the change in composition of particulate matter may be due to dilution and not to chemical gain or loss on the stream-originating particles alone. This is a difficult thing to be certain of, since the mineralogy in a strict sense will not be different over a broad region of a shelf area, thus confounding the identification of sources. More importantly, the metal-bearing phases cannot be diagnostically identified; therefore, a mixing curve of open-ocean particles poor in such phases with streamborne particles rich in them will not be visible mineralogically. Indeed the *definition* of the mixing may well be the distribution of trace metals on the particles. It is clear that analysis of particles for metals alone will not provide a singular answer to this problem.

Our study of the encounter of the Housatonic River with Long Island Sound at one time seemed to provide direct evidence for release of metals from particles at the boundary between freshwater and seawater. Both the Housatonic and the open waters of Long Island Sound showed much lower concentrations of cobalt, nickel, and silver than did the mouth of the Housatonic (Figure 12.3). Although these results might still be interpreted as showing release of metals, another explanation seems to be better.

The mouth of the Housatonic River is marked by a very large salt marsh area. The tidal range in this region is almost 2 m—the largest amplitude in Long Island Sound. In salt marshes as in marine deposits the storage of metals is related to indigenous reducing conditions. Sulfate in pore waters is reduced to sulfide, which is then sequestered by the ubiquitously available iron found on mineral surfaces. At low tide, the top layers of the marsh are aerated. At times of rainfall at low tide, especially, the sulfide phases in the marsh are oxidized and sulfate and associated metals are solubilized. As the tide comes in, the process is terminated and trapping of trace metals can occur again. Metal-rich particles from the streams are continuously trapped in the salt marsh, and atmospheric precipitation adds an additional burden. Thus, a concentration halo of certain trace metals in seawater is maintained around the salt marsh environs.

A similar process has been shown to occur on a horizontal scale in the Scheldt Estuary by Wollast (1975). There, reduced iron derived from the sediment is transported seaward with sulfide particles in anoxic waters. As the aerated open ocean water is encountered, the trace-metal-bearing sul-

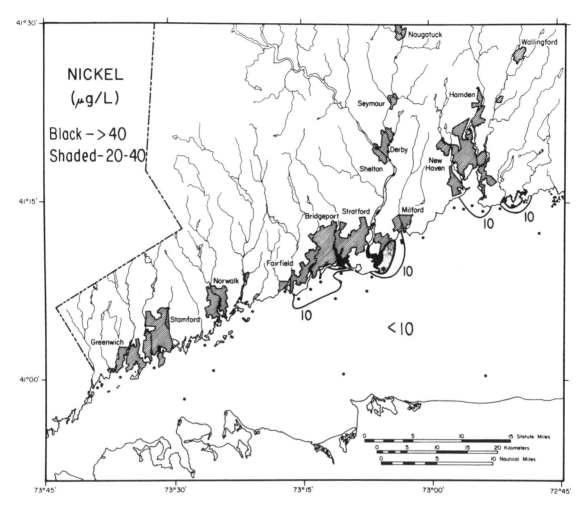

FIGURE 12.3 The nickel concentration increases sharply at the mouth of the Housatonic River. This is probably due to the release of metals from particles at the freshwater-seawater interface. The mechanism may be enhanced, if not actually controlled, by the presence of salt marshes.

fide phases are oxidized and a marked increase in dissolved copper concentration occurs simultaneously with a sharp decrease in dissolved iron as oxidation and precipitation of the oxide occur. The pattern of trace-metal release or precipitation occurring at a river mouth thus can be a complicated one.

We must discuss one additional factor influencing the fate of metals at the stream–estuary boundary. That is the hold-up time at the encounter. The stream encountering the sea is essentially ponded to some degree before it overflows or mixes with its saline barrier. During its holdup, reactions typical of reservoir situations can occur. Metals as well as silicon are removed from the water column, as is seen in the Connecticut River (Figure 12.4). After the water leaves the system by breaching or mixing with the salty estuarine barrier, it essentially mixes conservatively with the ambient estuarine waters. The influence of the tides appears to be paramount in providing the mechanism for holding up the stream water. It is not unreasonable to expect that in stream systems experiencing small tidal effects the trace metal concentration patterns during mixing are essentially conservative.

WHAT HAPPENS IN THE LARGER ESTUARINE MIXING BASIN

The large estuarine system is an important arena for further modification of the water before it becomes a part of the open ocean system. Such systems as Long Island Sound, Chesapeake Bay, and the Baltic Sea, although different in many ways, share the common property of being very large mixing basins. A mixing basin is subject to many of the same effects controlling the fate of trace metals as streams and the stream–seawater encounter discussed above. The extent of the modification depends on the length of time distinctively estuarine processes have to act on the water column. A fast flushing rate (i.e., completed in days), essen-

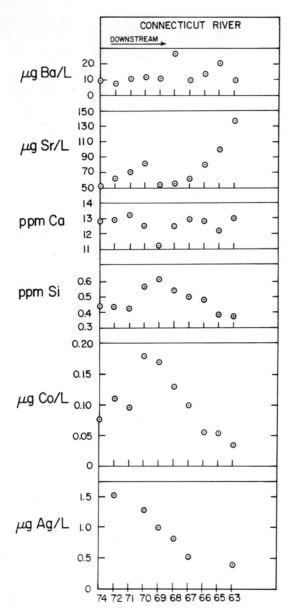

FIGURE 12.4 Both metals and silicon decrease in concentration in the freshwater, tidally affected parts of the Connecticut River. This can be ascribed to a reservoir effect based on width of the river mouth and tidal cycling.

tially makes the estuary a conduit; a slow flushing rate (completed in weeks) makes it a standing body of water with special properties due to the presence of seawater and tides.

It is the combination of biological activity in the sediment column and physical movement of the waters by storms and tidal action that makes the estuarine basin a particularly active region chemically.

By using ^{210}Pb from atmospheric precipitation primarily and ^{234}Th from *in situ* production from dissolved ^{238}U as

tracers, it is possible to identify possible mechanisms capable of modifying estuarine water composition. The work on ^{234}Th in Long Island Sound by Aller and Cochran (1976) clearly shows that its residence time in the water column is about one day. This is the same order of magnitude of time as that for ^{210}Pb in streams. The fact that no dissolved ^{210}Pb can be identified in Long Island Sound (Figure 12.5) implies that it too is rapidly removed in estuarine systems, resembling ^{234}Th in behavior. We will assume that this is indeed true and search for an appropriate mechanism.

It can be shown on the basis of material balance calculations that plankton cannot be important in transferring ^{210}Pb (and presumably ^{234}Th) to the estuarine floor. It does not seem likely that clay minerals are important scavengers since in the more propitious freshwater system they do not seem to be very effective agents.

The most likely agents appear to be the manganese and iron released from the reducing sediments as they are oxidized in the water column and form fresh precipitates. The precipitates deposit on suspended particles and there act as scavengers for a large number of trace metals. On return to the ocean floor by settling, the particles are reworked into the sediment by burrowing organisms. When the manganese and iron are recycled by the reduction-release-oxidation-precipitation steps, the trace metals carried down by the process mainly remain trapped in the sediment column, as seen by the material balance calculation for ^{210}Pb. Only where trace-metal-bearing sulfide particles are oxidized in the water column is there a refluxing of the associated metals to the waters again.

By this process, an extended residence of water in an estuarine basin results in a scavenging of the metals from the water column into the sediments.

Some of the finest grained manganese and iron oxide particles, with their associated trace metals, will be swept out to sea. This seems to be verified by the material balance

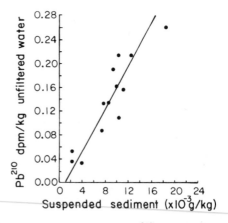

FIGURE 12.5 A plot of the total ^{210}Pb activity in Long Island Sound water versus the amount of suspended sediment indicating virtually no dissolved ^{210}Pb in the Sound.

studies on man-made [55]Fe reported by Labeyrie *et al.* (1975). The deficiency in some coastal sediments may be due to loss to the open ocean from the estuarine system. This manganese and iron could be a prime source of supply for the ubiquitous ferromanganese nodules in deep-ocean deposits.

REFERENCES

Aller, R. C., and J. K. Cochran (1976). [234]Th/[238]U disequilibrium in near-shore sediment: Particle reworking and diagenetic time scales, *Earth Planet. Sci. Lett. 29,* 37.

Benninger, L. K. (1976). The uranium-series radionuclides as tracers of geochemical processes in Long Island Sound, Ph.D. Thesis, Yale U., New Haven, Conn.

Benninger, L. K., D. M. Lewis, and K. K. Turekian (1975). The use of natural Pb-210 as a heavy metal tracer in the river–estuarine system, in *Marine Chemistry in the Coastal Environment,* Symp. Series 18, Am. Chem. Soc., Washington, D.C., pp. 202–210.

Boyle, E. A. (1976). The marine geochemistry of trace metals, Ph.D. Thesis, MIT, Cambridge, Mass.

Holtzman, R. D. (1964). In *The Natural Radiation Environment,* J. A. S. Adams and M. Lowder, eds., U. of Chicago Press, Chicago, Ill.

Kharkar, D. P., K. K. Turekian, and K. K. Bertine (1968). Stream supply of dissolved silver, molybdenum, antimony, selenium, chromium, cobalt, rubidium and cesium to the oceans, *Geochim. Cosmochim. Acta 32,* 285.

Labeyrie, L. D., H. D. Livingston, and V. T. Bowen (1975). Comparison of the distributions in marine sediments of the fallout derived nuclides [55]Fe and [239,240]Pu: A new approach to the chemistry of environmental radionuclides, in *Proceedings ERDA/ IAEA International symposium on transuranium nuclides in the environment,* San Francisco, Calif.

Lewis, D. M. (1976). The geochemistry of manganese, iron, uranium, lead-210 and major ions in the Susquehanna River, Ph.D. Thesis, Yale U., New Haven, Conn.

Martin, J. M. (1971). Contribution à l'étude des apports terrizenes d'oligoelements stable et radioactifs à l'ocean, Ph.D. Thesis, U. of Paris, Paris, France.

Sholkovitz, E. R. (1976). Flocculation of dissolved organic and inorganic matter during the mixing of river water and seawater, *Geochim. Cosmochim. Acta 40,* 831.

Turekian, K. K. (1971). Rivers, tributaries and estuaries, in *Impingement of Man on the Oceans,* John Wiley and Sons, New York, pp. 9–73.

Wollast, R. (1975). Paper presented at the International Union of Geology and Geophysics meeting, Grenoble, France, Sept.